SECOND EDITION

EXPLORING STATISTICS WITH THE IBM® PC

David P. Doane
Oakland University

Addison-Wesley Publishing Company

Reading, Massachusetts ■ Menlo Park, California ■ New York
Don Mills, Ontario ■ Wokingham, England ■ Amsterdam ■ Bonn
Sydney ■ Singapore ■ Tokyo ■ Madrid ■ Bogotá ■ Santiago ■ San Juan

Library of Congress Cataloging-in-Publication Data

Doane, David P.
    Exploring statistics with the IBM PC.

    Includes index.
    1. Statistics—Data processing.   2. Mathematical
statistics—Data processing.   3. IBM Personal Computer—
Programming.   I. Title.
    QA276.4.D63   1988   519.5′028′5526   87-1844
    ISBN 0-201-11821-1

It is a violation of copyright law to make a copy of the accompanying software, except for backup purposes to guard against accidental loss or damage. Addison-Wesley assumes no responsibility for errors arising from duplication or modification of the original programs.

Addison-Wesley Publishing Company, Inc., makes no representations, express or implied, with respect to this documentation or the software it describes, including without limitations, any implied warranties of merchantability or fitness for a particular purpose, all of which are expressly disclaimed. Addison-Wesley, its distributors and dealers shall in no event be liable for any indirect, incidental or consequential damages. The exclusion of implied warranties is not permitted by some statutes. The above exclusion may therefore not apply to you. This warranty provides you with specific legal rights. There may be other rights that you have which may vary from state to state.

IBM is a registered trademark of International Business Machines Corporation.

Original disk format and SAVON/SAVOFF program utility by Paul Amaranth. Used by permission.

Tables printed on pages 288, 289, and 290 were drawn from E. S. Pearson and H. O. Hartley, *Biometrika Tables for Statisticians,* 3rd Edition (Cambridge University Press, 1970), pp. 171–173. Reproduced by permission of the Biometrika Trustees.

 **To Blythe**

Preface

Objectives of This Book

This book is written to serve as a companion to a wide range of statistics textbooks. Its aims are to help you make the transition toward relying on computers and to show you that computationally complex methods can be used routinely. The computer case studies are designed to illustrate concepts explained in your textbook.

Programs are organized alphabetically, rather than by topic, so you can choose whichever programs you need. An instructor may indicate the preferred order to use the programs. But as the title suggests, you are encouraged to explore. There is no lock-step method here. You can sneak ahead.

Although the writing style of this book assumes you are a mature learner, it doesn't require you to be a mathematician or computer programmer. This material is accessible to anyone with a command of simple algebra. Mathematical notation is minimal, to avoid conflict with your other textbook(s).

Statistics is a tool that can help you learn about the world around you. You can use the EXPLORE programs to look at relationships among variables, or to estimate parameters that may affect your life. To you as a learner, sheer volume of data is no longer very important. You can handle big, real-world data sets as well as small, artificial, textbook problems. For this reason, this book uses realistic data whenever possible. You will see case studies that actually describe the world in which you live. You'll have access to real-world databases. You'll begin to build skills in handling data, right from the beginning.

Why Is This Book Necessary?

This book is written to help you use the microcomputer to learn statistics. Now you can join the many well-trained people in the world who confidently rely on computers and statistics. Their secret is this: Computers make it *easy* to deal with data.

But surely statistics is not really an easy subject? This is obviously a relative question. Good statisticians make it *look easy*. But if statistics were as easy as listening to a radio, everyone would be an expert. Statistics is the interface between mathematics and the real world, and there are many difficulties to be mastered. That's one reason why good statisticians are always in demand, and will continue to be.

Consider an analogy: Almost anyone can drive a car; fewer can drive a truck; still fewer can pilot an airliner. Yet all of these tasks are done every day, and those who have acquired the skills make it look easy because their skills have been honed by practice and serious study. The same is true of statistics: Statisticians aren't born with their skills.

Those who can effectively use computers, as well as statistics, are going to be secure in our modern technical world. Like the highly trained pilot, they will often enjoy unique career possibilities. But the analogy is imperfect, because statistics is more than a skill: It is a way of approaching problems and of structuring our thinking.

Computers are more than mere computational devices to aid statisticians. Naturally, we want to let computers handle the calculations. But modern computers are so powerful that they permit previously unknown ways of looking at the world. Computer-intensive statistics is still a relatively new field, and it's too early to say just what its impact will be on statistical science—but it will be significant.

Computers are not an unmixed blessing. One thing you may learn about computers is "reasoned skepticism." With practice you'll begin to see what can go wrong and why. You'll learn from the errors you make. Like any resource, we should only use computers up to the point where the marginal costs begin to exceed the marginal benefits. We want computers to become part of the *solution* to problems facing modern organizations, not part of the problem.

■ Why Study Statistics?

Why is statistical training so important for anyone who works in today's world? Two reasons: too much information or too little information. If insufficient data are available, statistical sampling can be used to obtain what's needed. If information must be extracted from large databases, statistical sampling can do the job scientifically. Most large organizations are, in fact, closer to drowning in data than to starving for data. Their problem is often not lack of data, but sifting massive piles of numbers for *relevant* information.

Statistics can help summarize large amounts of data and reveal relationships that are invisible just from looking at stacks of numbers. Statistics can yield decision rules to assist in operating a business. Statistics can squeeze the maximum information from small samples. Statistics can guide decision-makers in choosing samples of the right size. Statistics can ensure scientific sample selection. Statistics can help spot trends and present simple conclusions. Any way you look at it, statistics increases the ratio of *usable* information available to decision-makers.

Modern organizations need employees who aren't afraid of numbers. Although possession of statistical skill confers power in today's world, this is not to say that you should become a full-fledged statistician. Few organizations have regular access to professional statisticians. That's okay, since not every problem really requires a professional statistician. But in our technical world, everyone has to handle garden-variety statistical problems.

Someone once said that the main thing you need to know about statisticians is when to call for one. There's a lot of sense in this statement. You should learn enough about statistics to know when you've reached the limits of your knowledge or when you're on shaky ground. Naturally, if you are a scientist or an engineer (or if you work with them), you will have a need for above-average statistical skills.

■ Role of Microcomputers

We're all familiar with the mainframe data processing computers that large organizations use to handle their financial and accounting functions, payroll, benefits, and production planning and control. Mainframe computers have been used for decades. They are intended for large-scale data handling, information processing, storage, retrieval, and backup.

Microcomputers are relatively new on the scene. But they are key ingredients in any organization's data management strategy. For the first time in history, many individuals in an organization can enjoy hands-on access to significant desk-top computing power and mass storage of information. These micros may be more powerful than some mainframes, and they are cheaper. The implications for organization of work are profound.

With no programming skills, you can carry out any statistical calculations you desire. In conjunction with a database management program, your statistical package can analyze any subset of data available to your microcomputer. In short, you possess computational power beyond the dreams of statisticians only a few decades ago. You will take such power for granted and use it routinely.

In large organizations, microcomputers are mostly used to manipulate subsets of data stored on larger computers or to create files that can be reintegrated into larger databases. In smaller organizations, or for departments within larger ones, microcomputers are the preferred mode of operations, because of the convenient hands-on access they provide to data and programs. Networking further enhances the appeal of micros. However, advances in telecommunications may favor a continued strong role for mainframe time sharing.

The main advantage of personal computers to you right now is that you can carry your EXPLORE computer programs home or take them to work—

anywhere you have access to a microcomputer, you can sit down and do statistical calculations. Never before has this been possible. College students who learned to use computer packages used to leave the programs behind when they graduated. Now, they can take the programs with them!

The organization in which you work will have various statistical packages available. They will perhaps be larger or more powerful than the EXPLORE programs, but the idea is the same. In the future, you can expect to see more exotic statistical tools on microcomputers: higher speed computation, vivid color graphics, massive data handling, and true integration with database management. You will have no trouble adapting to these possibilities, once you've oriented yourself using EXPLORE.

Changes in the Second Edition

The primary change in this second edition is the release of new software. However, the programs keep the same look and feel, and individual changes are mostly internal.

Four new programs have been added: GROUP (calculations for grouped data); SAMPLER (repeated random sampling from a file); SAMSIZE (sample size for specified precision); and SPLIT (sample splitting on a related file).

We have improved the performance on hard disk systems. Programs can handle drives C and D, and the diskettes are not copy-protected. The text now includes hard disk instructions, and the file EXPL.BAT assists hard disk users.

We have improved the trapping of odd user inputs and error handling, refined the column and table formats, increased the accuracy of probability calculations, enlarged some data arrays, and added new program options. In ANALYZ, BIGRES, and MGRES we have provided a case label option (to add the names of states, for example). Other new features include a new 360K diskette format, to permit DOS 2.0 through 3.1 to be installed directly on the EXPLORE program diskette. The accompanying database (DATABASE-1) has been updated with time series files that cover 1960–1985. A new database (DATABASE-2), with files that include nations and metropolitan areas, is available separately from the Addison-Wesley marketing manager.

A new "screen capture" program will save screens in ASCII files, for subsequent word processing or printing (e.g., on a network). SAVON reassigns the PrtSc key to print to a file; SAVOFF restores the normal PrtSc function. Chapter numbers have been added in the manual, for reference purposes. Finally, we have included an extra copy of the help file (PINFO.EXE) on the data disk; users can safely eliminate this file from the program disk in order to accommodate larger versions of DOS.

Equipment Needs

To run the software accompanying this book you will need the following equipment as a minimum:

IBM PC (or compatible)

64K memory

One double-sided disk drive (two are recommended)

Monochrome or color monitor

Printer (optional)

DOS (version 2.0 or later) system disk

Additional systems information appears in Chapter 1.

■ Acknowledgments

Two individuals deserve special recognition. Paul Amaranth of the Office of Computer Services, Oakland University, gave expert technical guidance on mainframe editing, computational algorithms for statistical tables, communications protocols, and about microcomputers in general as this project unfolded. Gerald Post of Oakland University not only provided an improved regression algorithm but also cheerfully answered lots of technical questions and helped adapt software to support this project, often on short notice.

I would also like to thank Kenneth C. Young of Eastern Michigan University and Harvey A. Shapiro of Oakland University, who contributed to early versions of some of the computer programs explained in this book. Mike Langstrom of Volkswagen of America provided technical assistance on algorithms for finding areas under probability density functions. Edward Vesely of Kidder, Peabody Inc. supplied stock market data. Joe Henninger of Inacomp, Inc., was generous with helpful advice. Mary Coffey, Jerrold Grossman, Scott Monroe, and Ron Tracy of Oakland University offered suggestions and technical advice.

For carefully reviewing my programs and draft manuscripts, and for offering a host of suggestions that are embodied in the finished product, I wish to express my gratitude to Donald Adolphson of Brigham Young University, Alan Fask of Fairleigh Dickinson, Michael Hand of Willamette University, Steven Nahmias of the University of Santa Clara, Pamela Specht of the University of Nebraska, Paul Mason of the University of North Florida, and Robert Meier of Eastern Illinois University. Joseph A. Nordstrom of the University of Houston not only reviewed the manuscript but also provided extensive suggestions for using the programs in the classroom. I would also like to thank Gary Simon of New York University for his thoughtful comments.

Thanks are also due to the many students, especially Lori Marsee, Harash Sachdev, Roxanne Giordano, Kathleen Long, Steve Raimi, and James Abramczyk who have participated in the classroom testing of these computer programs, and whose suggestions have made them more workable. Responsibility for any remaining errors is my own, of course.

Rochester, Michigan   D.P.D

Contents

1

Getting Started

Diskette Overview

The EXPLORE package consists of two diskettes: the EXPLORE program diskette (which contains programs) and the DATABASE-1 diskette (which contains mostly data but also a few programs). Most of the programs are designed to analyze data. You will use the EXPLORE programs either in a dual floppy drive PC (personal computer) or with a hard disk PC. In either case, you should have some idea of what is stored on the two diskettes.

■ Contents of Program Diskette

A list of files stored on the EXPLORE program diskette is shown in Table 1.1. Computer-oriented readers will be able to guess what these files are. Others need not worry about it. However, to give you some idea of storage requirements, approximate file sizes are given (K = kilobytes). Hard disk users will wish to transfer these files to their hard disk, as explained later in this chapter. We will discuss the programs shortly. But first let's take a look at the DATABASE-1 diskette.

■ Contents of DATABASE-1 Diskette

The DATABASE-1 diskette contains eighty-two data files, a help program to describe the data files, and a duplicate copy of the main help program (the last file in Table 1.1). The help programs are invoked by the EXPLORE main menus, which are explained in Chapter 2. The eighty-two data files are in ASCII format. Table 1.2 gives a general description of the DATABASE-1 diskette contents, along with approximate file sizes.

The files shown in Table 1.2 do not fill the DATABASE-1 diskette. There is room for quite a few of your own workfiles. Detailed descriptions of the DATABASE-1 files will be given shortly or may be found by using the help command explained in Chapter 2. You can also look at Appendixes H and I to see a listing of the files.

Table 1.1   Contents of EXPLORE Program Diskette

Description	File Name	Size
Basic runtime module	BASRUN.EXE	32 K
EXPLORE main menu	EXPLORE.EXE	4 K
Descriptive statistics	ANALYZ.EXE	14 K
Analysis of variance	ANOVA.EXE	11 K
Areas under curves	AREA.EXE	8 K
Bivariate regression	BIGRES.EXE	12 K
Binomial probabilities	BINOM.EXE	6 K
Box plot	BOXPLOT.EXE	7 K
Chi-square independence	CHI.EXE	14 K
Simulated dice rolls	DICE.EXE	9 K
Goodness-of-fit tests	GOODFT.EXE	18 K
Grouped data analysis	GROUP.EXE	8 K
Formatted file display	LAYOUT.EXE	7 K
Matrix of correlations	MATCOR.EXE	6 K
Multiple regression	MGRES.EXE	18 K
Central Limit Theorem	MONTE.EXE	9 K
Scatter plot	PLOT.EXE	8 K
Poisson probabilities	POIS.EXE	5 K
Random number generator	RAND.EXE	8 K
Sampling from a file	SAMPLER.EXE	7 K
Sample size estimation	SAMSIZE.EXE	5 K
Sorted file display	SORTER.EXE	10 K
Split a file	SPLIT.EXE	9 K
Transform variables	TRANS.EXE	9 K
Time-series trends	TREND.EXE	15 K
Two sample tests	TWOSAM.EXE	9 K
EXPLORE file editor	DFILE.EXE	14 K
Program initiation	START.EXE	3 K
File save on	SAVON.COM	4 K
File save off	SAVOFF.COM	1 K
Automatic booting	AUTOEXEC.BAT	1 K
Hard disk sample setup	EXPL.BAT	1 K
General program help	PINFO.EXE	24 K
Reserved for DOS transfer		40 K
Unallocable space		4 K
Free disk space		3 K
Total		362 K

Table 1.2   Contents of DATABASE-1 Diskette

Description	File Name	Size
Database information	DINFO.EXE	8 K
General program help	PINFO.EXE	24 K
State database		45 K
Forty-four data files, each containing fifty observations representing the various individual states in the United States.		
Time-series database		40 K
Thirty-eight data files, each containing twenty-six observations on an economic time series, representing the years 1960–1985.		
Available space		245 K
Total		362 K

■ Using a PC with Dual Floppy Disk Drives

The general idea is that the EXPLORE program diskette goes in drive A and the DATABASE-1 diskette goes in drive B, as shown in Figure 1.1. On some PCs the drives may be arranged top-and-bottom rather than side-by-side.

The EXPLORE program diskette does not contain DOS (the disk operating system) which is needed to boot the PC when you turn it on. A diskette

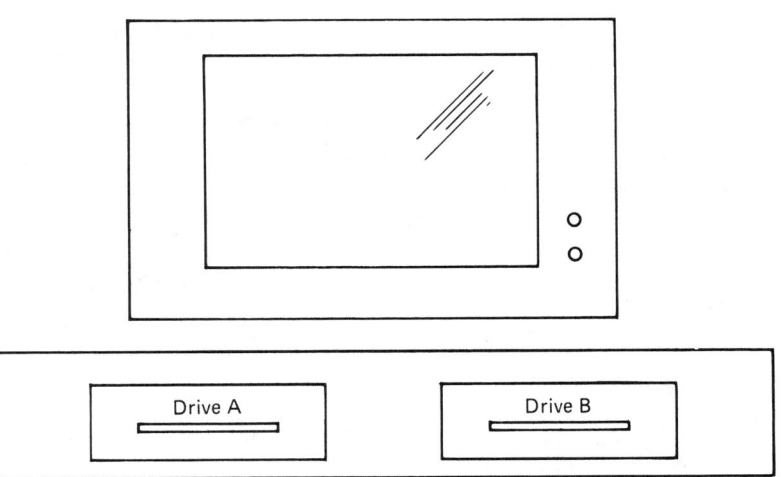

Figure 1.1   Personal Computer with Two Disk Drives

containing DOS is *self-booting*. The advantage of a self-booting diskette is that you do not have to insert a separate DOS diskette every time you turn the PC on or when you exit from the EXPLORE programs.

It would be nice if we could have included DOS on the EXPLORE program diskette. But DOS is a proprietary product that varies among computers, and we have no way of knowing which version you are using. However, we have reserved space on the EXPLORE program diskette which will allow you to transfer some versions of DOS.

Rather than transferring DOS to your original EXPLORE program diskette, it is probably a better idea to keep it in a safe place, and instead make a new self-booting program diskette that contains your version of DOS and the EXPLORE programs. If you prefer to make a new self-booting diskette, skip the next section.

Transferring DOS to EXPLORE Program Diskette

If you transfer DOS to your original program diskette, there may be little or no room left for data files. Normally, data files would go on a different diskette anyway, but you should be aware of the situation before you start. If you wish to transfer DOS to your original program diskette, place your DOS diskette in drive A and turn on the computer, if it is not already on. You should see the prompt A >. If you have a choice, use DOS 2.0 or 2.1 rather than DOS 3.0 or above to allow more room for programs or data. Place your EXPLORE program diskette in drive B and type

SYS B:

The computer should report that the (hidden) system files have been transferred to the EXPLORE program diskette. Then type

COPY COMMAND.COM B:

to copy the DOS command file to the EXPLORE program diskette. If everything works, you will have a self-booting diskette with little or no room left for data files.

If you get an error message "Insufficient Room" (which will happen with DOS 3.0 or higher) you can free up the necessary space to copy the command file by erasing the help file PINFO.EXE. You will still have the help file PINFO.EXE on the DATABASE-1 diskette, but it will not be available if you are using your own data diskette. Probably you will rarely need the help file by the time you start using your own data diskette anyway. If these tradeoffs are acceptable to you, type

ERASE B:PINFO.EXE
COPY COMMAND.COM B:

If these steps are successful, skip the next section. Otherwise, check your steps for errors and try again, or use the procedure described in the next section.

Making a New Self-Booting Diskette

Making a new self-booting diskette is easy. All you need to do is format a new blank diskette with your version of DOS and copy the EXPLORE programs onto the new diskette. The EXPLORE diskettes are not copy-protected as a convenience for those who purchase the programs (not, of course, as an invitation to make illegal copies to avoid paying for this copyrighted, modestly priced software). If you don't want a self-booting diskette, skip to step eight. Otherwise, follow all steps carefully.

1. Obtain a new blank diskette. (You can buy them singly or in boxes of ten in most college bookstores or office supply stores or from student organizations at many universities.) Place the new blank diskette in drive B.

2. Place your DOS diskette in drive A and turn on the computer if it is not already on. If you have a choice, using DOS 2.0 or 2.1 will leave more room on your new diskette for programs or data.

3. When you see the A> prompt, verify that the disk drive door is closed on drive B and that it contains your new *blank* diskette (otherwise, whatever is on the diskette in drive B will be wiped out). Then type

    FORMAT B: /S

    The computer should format the new blank diskette and install DOS on it. If you get an error message (such as "file not found"), check your DOS diskette to be sure it contains the DOS file FORMAT.COM. If not, locate a DOS diskette that does contain FORMAT.COM.

4. Remove the DOS diskette from drive A and insert the EXPLORE program diskette. Leave your newly formatted blank diskette in drive B and type

    COPY A:*.* B:

    The computer should copy 33 files onto your new diskette. *This will take a while. Do not open the drive doors or try to remove the diskettes until the red drive lights are off.* Unfortunately, with some versions of DOS there will not be room for the last file PINFO.EXE. If not, do not worry about it—a duplicate copy is already on the DATABASE-1 diskette, where it can be accessed by the programs.

5. Remove the EXPLORE program diskette from drive A, and put it in a safe place. You may need it in case you wipe out your programs by accident some day.

6. Remove your newly created diskette from drive B and put a label on it saying "EXPLORE Program Diskette." It is a good idea to write your name on it too, in case you forget it in a PC drive some day.

7. Place your newly created EXPLORE program diskette in drive A, and

place the DATABASE-1 diskette in drive B. To verify that the new diskette is self-booting, simultaneously press the

Ctrl-Alt-Del

keys on your PC (all three keys at once). The opening screen should appear. If not, ask for help or go back to step 1.

8. If your PC has already been booted, you may start the EXPLORE programs without turning off the PC. Make sure that the program diskette is in drive A and the DATABASE-1 diskette is in drive B, and type

START

If a program crashes, you may restart by typing START (or by typing EXPLORE, or by typing the name of a particular program, as explained under "Some Special Situations" at the end of this chapter).

After that, all programs are menu-driven. The computer will read files from the diskettes, searching both drives until it finds what it is looking for. If anything goes wrong and you end up back in the main operating system (as evidenced by the A> prompt), you may start over by typing START, and everything should be back to normal. You may type all commands in lowercase letters (they are printed here as uppercase letters only for clarity).

The next time you use EXPLORE, insert your self-booting EXPLORE program diskette in drive A and insert the DATABASE-1 diskette in drive B. Then turn on the computer. The EXPLORE menus should appear automatically. If not, type START, and everything should be fine.

If you wish, you may use your own data diskette instead of the DATABASE-1 diskette. All the statistical programs will work normally. However, the "help" program will not be available unless you were able to copy the file PINFO.EXE from the original EXPLORE program diskette onto your self-booting EXPLORE program diskette, as explained in step 4 above. In that case you may still copy PINFO.EXE onto your own data diskette if you wish.

■ What about Hard Disk Users?

The EXPLORE programs are not copy-protected in this second edition, so you can copy them onto your hard disk. At the end of this chapter, detailed instructions are given for installing and using EXPLORE programs on a hard disk.

■ Getting Printed Copy

Naturally, you will often want to print a particular screen for future reference. If you have a printer attached (and turned on), you can use the PrtSc key on your IBM PC. Other computers may have similar procedures for printing a particular screen. If not, or if you do not have a printer, just take notes from the screen as necessary.

The EXPLORE programs are set up to make it easier to print a single screen at a time. The computer often pauses to issue the message "Press any key to continue." These pauses are inserted to give you time to use the PrtSc key to print a screen before proceeding. The PrtSc key will not cause the computer to proceed to the next screen. When the printing task is finished, you may proceed to the next screen by pressing some other key.

Why don't the EXPLORE programs just send everything to the printer? Partly because you will not want to print everything. The screen-at-a-time approach lets you print only what is really important. Moreover, printers vary greatly, and it would be difficult to anticipate exactly how yours works (or even if you have a printer at all).

A few EXPLORE programs optionally use special graphics characters, which your printer may not print. If you plan to print screens using the PrtSc key, you might prefer not to use these optional graphics characters, even though they will look nice on the screen. A little experimentation will not hurt anything if you are not sure what your printer will do. Use of graphics characters is kept to a minimum and is kept optional to ensure that you will not run into unnecessary problems.

■ Saving Screens to Disk

Instead of printing them, you can save each screen to disk, for word processing or later printing, by redirecting the PrtSc key with the SAVON program (contained on the program disk). For example, to save screens to a file on drive B named BOB1.DOC, *before* you start EXPLORE (while in DOS), type

SAVON B:BOB1.DOC

Each time you press PrtSc, another screen will be saved at the end of file B:BOB1.DOC (which quickly becomes large, so be sure your disk has enough room). If you do not specify a file name, SAVON will prompt you for one.

If you use SAVON, do not use the graphics options in some programs (most word processors cannot handle non-ASCII characters).

If you redirect PrtSc output to a file using SAVON, you must restore normal PrtSc function when you finish EXPLORE (and are back in DOS) by typing

SAVOFF

Forgetting to type SAVOFF may confuse you later (or someone else who uses the PC), since the PrtSc key remains redirected until the computer is turned off or SAVOFF is executed.

■ State Database

The *state database* stored on the DATABASE-1 diskette consists of forty-four data files, each containing actual data for the fifty states of the United States. Each file refers to a particular point in time. Table 1.3 lists definitions of the state data files that are already on the data diskette. Appendix H lists the contents of the state database in detail.

The state data files are an example of *cross-sectional data*. Cross-sectional data represent a group of individual units, such as persons, firms, or geographical regions. These data can be the result of a survey or may be obtained from official records such as the U.S. Census. It is a common statistical task to compare geographical regions such as cities, states, or countries.

Notice that all file names must be short (eight characters or fewer). One file (STATE) contains the actual name of each state to use as a label file in the EXPLORE programs to identify cases. Using Appendix H, you can identify states with particular characteristics. For example, which states have high

Table 1.3   State Data Files in DATABASE-1 (N = 50 States)

ABORT	= 1980 abortions per 1000 live births
AFDC	= 1980 average aid for dependent children per family
AGE	= 1980 median age of population
BEDS	= 1979 hospital beds per 100,000 population
BIRTH	= 1980 live birth rate per 1000 population
BLACK	= 1980 percent black population in state
CANCER	= 1980 cancer death rate per 100,000 population
CARS	= 1980 car registrations per 1000 population
COLLEGE	= 1980 percent college graduates among age 25 and over
DEBT	= 1980 state debt per capita
DEFENSE	= 1979 Defense Department spending per capita
DIVORCE	= 1980 divorce rate per 1000 population
DOCS	= 1980 doctors per 100,000 population
EDUC	= 1979–1980 percent of personal income spent on K1–K12
FEMLAB	= 1980 female labor force participation rate
INCOME	= 1980 per capita personal income
HEART	= 1980 heart disease death rate per 100,000 population
HOSP	= 1980 average hospital cost per day
INFMOR	= 1978 infant mortality rate per 1000 live births
LIFER	= 1980 percent of population that has always lived in state
NEAST	= 1 if state is in Northeast, 0 otherwise
OVR65	= 1980 percent of population over age 65

Source: U.S. Bureau of the Census, *Statistical Abstract of the United States, 1982–83*

divorce rates? Which have low divorce rates? Statistical tasks such as this will be greatly facilitated once you have learned to use the EXPLORE programs for data analysis.

The state database covers a wide range of social and economic indicators that will permit you to experiment and explore your own hypotheses. For example, it is likely that you will want to explore associations among these files. Using regression (program MGRES), you could study the relationships between income, unemployment rates, and urbanization and divorce rates:

$$DIVORCE = f(INCOME, UNEM, URBAN)$$

It is easy to see the importance of estimating such relationships. Of course, regression is only one of many statistical tools that you can use to explore the databases (and a fairly advanced one unless you have prior statistical training).

Even if cross-sectional data files do not cover exactly the same year, it is still possible to study relationships between them if the pattern is believed to be fairly constant over time. For example, even though the overall level of

Table 1.3   (continued)

PCRIME = 1980 burglary rate per 100,000 population
POP = 1980 population in millions
POPCH = 1970–1980 percent population change
POPDEN = 1980 population density per square mile
PUBAID = 1980 public aid recipients as percent of population
PUPIL = 1979–1980 pupil-to-teacher ratio
REPUB = 1980 percent of presidential vote going to Reagan
SALETX = 1980 retail sales tax in percent
SPANISH = 1980 percent Spanish origin population
SPEND = 1980 state government expenditures per capita
STATE = state name (abbreviated)
SEAST = 1 if state is in Southeast, 0 otherwise
TDEATH = 1980 traffic death rate per 100,000 population
UNBEN = 1980 average duration of state unemployment benefits
UND25 = 1980 percent of licensed drivers under age 25
UNEM = 1980 percent unemployment rate
UNION = 1980 percent union members, nonagricultural labor
URBAN = 1980 percent of population in urban areas
VCRIME = 1980 aggravated assaults per 100,000 population
VOTE = 1980 percent eligible voters actually voting
WAIT = 1 if at least two-day wait for marriage, 0 otherwise
WEST = 1 if state is in the far West, 0 otherwise

(Washington, D.C.: U.S. Government Printing Office, 1983).

unemployment may change from year to year, states tend to remain in the same *relative* positions from one year to the next. Not all the files in the state database refer to exactly the same point in time, owing to lags in reporting and publication, but they are probably close enough.

A few of the state files contain binary variables. A binary variable has only two possible values. We assign $X = 1$ if a certain characteristic exists and $X = 0$ otherwise. In this way we can include effects that are difficult to quantify. You can easily create your own new binary variables, or you can use those already in the database. For example, what effect (if any) does region have on divorce rates? The model could be proposed as

DIVORCE = f(INCOME, UNEM, URBAN, NEAST, SEAST, WEST)

Binaries NEAST, SEAST, and WEST would indicate regional effects. If their predictive power turned out to be important, we would know that regional effects do exist. Such variables are easily handled by using regression (program MGRES), which will be explained later on.

Time-Series Database

The *time-series database* stored on the DATABASE-1 diskette consists of thirty-eight time-series variables describing the U.S. economy. One group of variables covers the national income and product accounts, such as GNP and related aggregate financial flows. Another group of variables represents labor, prices, and productivity. A third group represents monetary aggregates, such as the nation's money supply and its components. Detailed definitions are listed in Table 1.4.

The years covered are 1960–1985, which provide a reasonable representation of recent U.S. experience. Appendix I lists the complete contents of the time-series database. Notice that short variable names (eight characters or fewer—even single-character names such as C, I, G, X, and M) are used.

A time series is a numerical quantity measured *at a given point in time* (such as the quantities shown on an accounting balance sheet) or *over a period of time* (such as the quantities shown on an accounting income statement). Our interest focuses on how the data vary over time rather than over space or between individuals. A different set of statistical techniques must generally be used to study time-series data than those used to study cross-sectional data.

With time-series data you can study patterns of individual change in a single variable. For example, how fast is the money supply rising? How fast are prices changing? What is the average rate of growth in business productivity? We can also analyze the relationships among time-series variables. For example, what is the relationship between personal consumption expenditure, personal income, and interest rates? We might postulate a model such as

C = f(PI, PRIME)

Table 1.4   Time-Series Data Files in DATABASE-1 (N = 26 Years)

YEAR =	year (1960 through 1985)
GNP =	gross national product (billions)
C =	personal consumption expenditures (billions)
I =	gross private domestic investment (billions)
G =	government purchases of goods and services (billions)
X =	exports of goods and services (billions)
M =	imports of goods and services (billions)
X-M =	net exports (billions)
PI =	personal income (billions)
PT =	personal taxes (billions)
PS =	personal saving (billions)
DPI =	disposable personal income (billions)
APS =	average propensity to save (percent of PI)
DPI-CAP =	disposable personal income per capita (dollars)
DEFICIT =	federal deficit on NIPA basis (billions)
LABOR =	civilian labor force (millions)
UNEMPT =	civilian unemployment rate (percent of LABOR)
PR-MEN =	adult male labor force participation rate (percent)
PR-FEM =	adult female labor force participation rate (percent)
PROD =	index of hourly business productivity (1977=100)
FARM =	index of hourly farm productivity (1977=100)
COMP =	index of hourly business compensation (1977=100)
FRB =	Federal Reserve Board percent manufacturing capacity utilization
CPI =	consumer price index, all items (1967=100)
PPI =	producer price index, finished goods (1967= 100)
FPI =	farm price index, all farm products (1977=100)
CH-CPI =	change in consumer price index in last year (percent)
R-3MO =	mean yield on three-month treasury bills (percent)
R-BOND =	mean yield on Moody's Aaa bonds (percent)
PRIME =	prime rate charged by U.S. banks (percent)
DJIA =	Dow Jones Industrial Average of 30 stocks
M1 =	currency + demand and other check deposits (billions)
M2 =	M1 + savings and small time deposits, etc. (billions)
M3 =	M2 + large time deposits + term RP's, etc. (billions)
CH-M1 =	percent change in M1 over previous year
CDEBT =	consumer installment debt (billions)
FDEBT =	federal government debt (billions)
MDEBT =	mortgage debt by all holders (billions)

Source: *Economic Report of the President* (Washington, D.C.: U.S. Government Printing Office, 1986).

However, you should be careful of relying on simple time-series models such as this because it has been found that many time-series variables are highly correlated even when they may actually have little to do with one another in a cause-and-effect sense. This is partly due to the fact that most things have an upward trend over time. It may also be attributable to complex underlying common causes. Fairly sophisticated statistical procedures might be necessary to handle time-series variables effectively. However, the EXPLORE programs do include methods of analyzing time-series data.

Notice that one file (YEAR) simply represents the year in which the observation is taken. When you run certain EXPLORE programs, you will probably want to use YEAR as a label file to ensure easy identification of individual data points.

■ Order of Program Use

The EXPLORE programs may be used in whatever order you wish. The typical statistics textbook would follow a sequence something like that in Table 1.5. Within each category, programs are listed alphabetically and not necessarily in the order in which they would be used.

Table 1.5   Typical Order of EXPLORE Program Use

Data description:	ANALYZ	description of one variable
	BOXPLOT	box plot for one variable
	GROUP	grouped data analysis
General purpose:	LAYOUT	formatted file display
	SORTER	sorted file display
	RAND	random number generator
	TRANS	variable transformations
Probability:	AREA	areas under curves
	BINOM	binomial probabilities
	DICE	simulated dice rolls
	MONTE	Central Limit Theorem
	POIS	Poisson probabilities
Hypothesis tests:	ANOVA	analysis of variance
	CHI	chi-square independence test
	GOODFT	goodness-of-fit tests
	MATCOR	matrix of correlations
	SAMPLER	sample a file
	SAMSIZE	sample size
	SPLIT	split a file
	TWOSAM	two-sample tests
Regression:	BIGRES	bivariate regression
	MGRES	multivariate regression
	PLOT	bivariate scatter plot
Time series:	TREND	time-series trends

In addition to the programs listed in Table 1.5, the EXPLORE file editor is of considerable importance, right from the first time you use the EXPLORE programs. The same is true for the EXPLORE help menu, which you will probably encounter the first time you use the program diskette. However, the help menu and file editor are not mentioned by name because they are entirely menu-driven. The file editor and help menu will be explained in detail in the next chapter.

When you create your own data files, the EXPLORE programs expect you to use one-component data file names like GROWTH or SCORES. You will be warned by the computer if you attempt to save data under a file name that might wipe out one of the EXPLORE programs, one of the DATABASE-1 files, or any other file that you have already saved. But for safety you should absolutely avoid the two-component file names shown in Tables 1.1 and 1.2.

■ Installing EXPLORE on a Hard Disk

The EXPLORE programs are not copy-protected in this second edition, so you can copy them onto your hard disk. This task should be done *only once*. First, you should make a subdirectory named EXPLORE. Do this from the root directory (also called the DOS directory) by typing

MD EXPLORE   *(makes a new subdirectory named EXPLORE)*
CD EXPLORE   *(takes you to the EXPLORE subdirectory)*

Then you should place the EXPLORE program diskette in drive A and type

COPY A:*.* C: *(copies EXPLORE programs to hard disk drive C)*

This assumes that the EXPLORE program diskette is in drive A and your hard drive is named C (if not, use whatever letters you need). You could repeat this last step with the DATABASE-1 diskette, but you may not wish to copy the database files onto your hard disk (it will clutter your disk with 82 data files). If you do not copy the database files to the hard disk, you must still insert the database diskette in drive A or drive B when you run the programs.

Be sure the hard disk is the logged drive (automatic if you boot from the hard disk) before starting the EXPLORE programs, and *be sure that you execute a PATH command (if necessary) telling the computer where to find the DOS files it needs*. For example, you might have your DOS system files (including the important command ASSIGN.COM) in a subdirectory named C:\DOS, so your PATH command might suggest looking first in the root directory C:\ and then in the DOS subdirectory C:\DOS. In this example, the command would be

PATH C:\;C:\DOS

The PATH command could be inserted in your AUTOEXEC.BAT file, or it could be inserted in a special batch file named EXPL.BAT.

■ File Search on a Hard Disk

Most EXPLORE programs seek files first on the default drive (logged drive) followed by drive B and then drive A. It is assumed that drives A and B exist. If not, or if you do not plan to use them, or if you have an unusual drive such as D (possibly a virtual disk), you *must use the ASSIGN command before running EXPLORE, or you will encounter delays or puzzling error messages.* For example, if you have drives A and C but no drive B, before starting EXPLORE (while in DOS) you could type

ASSIGN B = C

or

ASSIGN B = A

These commands tell the computer that when a program wants to "look on drive B," it should look on another drive instead. If you have a drive B but never plan to use it, you could use the ASSIGN command to tell the computer which drive to use when drive B is requested. If you have a drive such as D (maybe a virtual disk), you could type

ASSIGN B = D

These examples are intended to be illustrative, not exhaustive. There are too many situations to anticipate them all. If you do not reassign unused drives A or B, at least keep a floppy disk in them to speed file searches (an empty drive slows things down a lot).

The DOS file ASSIGN.COM must be stored in the same directory from which you execute the ASSIGN command (or a PATH command must already have been executed to tell the computer where to find ASSIGN.COM).

If you use the ASSIGN command, it is best to do so in a special batch file, to be sure that you do not forget it. A sample batch file named EXPL.BAT is included on the program diskette. Local situations vary so much that it is impossible to include a batch file that will suit everyone's needs, but you can modify EXPL.BAT to fit your situation. This is what file EXPL.BAT looks like (explanations are given at the right):

Command	Effect
assign b = c	reassigns drive B to drive C
cd explore	takes you to EXPLORE directory
pause	inserts a pause
start	runs START.EXE to start EXPLORE
cd \	returns you to the root directory
assign b = b	reassigns drive B to itself

You could use the EDLIN editor to modify EXPL.BAT to suit your needs by deleting or adding lines. After editing the file EXPL.BAT to suit your needs, copy it to the root directory, and always type EXPL to invoke the EXPLORE programs to be sure that the correct commands are executed.

Assuming that the hard disk is your logged drive, when you are asked for a drive for saving data files, you may just press the Enter key (the default drive is assumed). You may still read files from a floppy drive A or B unless you have reassigned the drives.

■ Some Special Situations

Any EXPLORE program can be run directly from DOS just by typing its name (ANALYZ, for example). Advanced users may come to prefer this method to using the main menu as explained in Chapter 2. But hard disk users, if you by-pass EXPL.BAT, you must remember to use the ASSIGN command before leaving DOS if it is needed.

The EXPLORE programs run on many IBM-compatible PCs. However, you will have to investigate this for yourself if you have an IBM-compatible computer.

The EXPLORE diskettes and key programs are in unprintable (compiled) form. They utilize a proprietary IBM product (BASRUN.EXE), which cannot legally be copied or distributed except by written license. However, the instructor's manual does contain partial listings of some EXPLORE programs for those who need to know how things work. We have elected not to distribute source code for a variety of reasons, not the least being the complex steps that we have taken to compress and compile the programs. The process would be difficult to explain or duplicate without extensive consultation.

EXPLORE programs assume dual floppy disk drives. But if you have only one floppy disk drive, the EXPLORE programs can be made to work after a fashion if you are very patient, with frequent switching of diskettes. If the computer cannot find something, it will prompt you to insert the other diskette. The ideal situation is a hard disk, which materially speeds up the programs.

EXPLORE site licenses, for multi-users, can be purchased from the publisher. See page 304 for further details.

2

Data File Editing

Introduction

As we saw in the previous chapter, you are going to be using a collection of separate computer programs called the EXPLORE programs. These programs are written in BASIC and are stored in compiled form on diskettes. Each program can be run separately from DOS by typing its name (ANALYZ, for example). But a program usually begins when it is called by another program and ends when it calls some other program. The programs are organized so that several menus provide control over everything. The four main "places you can be" are shown schematically in Figure 2.1.

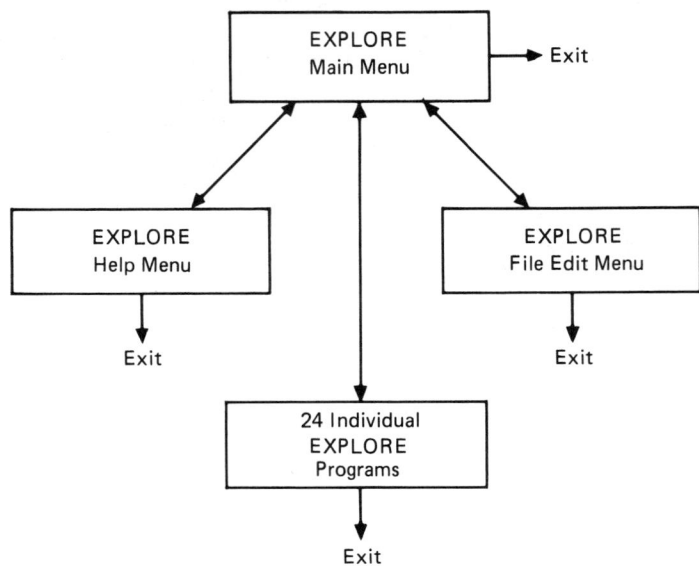

Figure 2.1   Organization of EXPLORE Programs

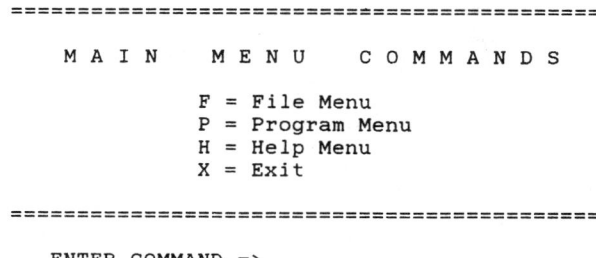

```
================================================
     M A I N    M E N U    C O M M A N D S
               F = File Menu
               P = Program Menu
               H = Help Menu
               X = Exit

================================================
        ENTER COMMAND =>
```

Figure 2.2   Main EXPLORE Menu

The *main menu* provides master control over everything. Two-way communication exists between the main menu and all the other programs. The main menu can call any of the other programs, and they in turn can call back the main menu. But you always have the option to *exit* from any program (that is, exit entirely from the EXPLORE programs). If you exit, you will go back to DOS, as evidenced by the A> (or C>) prompt.

■ Overview of Menus

When you run the master control EXPLORE program, the main menu will be presented. The function of this menu is to call the various other programs. The main menu looks like Figure 2.2.

Your command choice depends on what you are doing. Usually, you just want to run a particular program, so you type P. The *program menu* shown in Figure 2.3 will then appear. From this menu you can call any of the 24 individual programs.

```
---------------------------------------------------------
|     E X P L O R E    P R O G R A M    M E N U         |
|                                                       |
|            S e c o n d    E d i t i o n               |
|                                                       |
|=======================================================|
|                                                       |
|   1 ANALYZ      7 CHI       13 MGRES      19 SAMSIZE   |
|   2 ANOVA       8 DICE      14 MONTE      20 SORTER    |
|   3 AREA        9 GOODFT    15 PLOT       21 SPLIT     |
|   4 BIGRES     10 GROUP     16 POIS       22 TRANS     |
|   5 BINOM      11 LAYOUT    17 RAND       23 TREND     |
|   6 BOXPLOT    12 MATCOR    18 SAMPLER    24 TWOSAM    |
---------------------------------------------------------
Enter number of program you want (press ENTER to quit)?
```

Figure 2.3   Program Menu

```
====================================================
FILE    EDIT    COMMAND    MENU

        C = create a new file
        D = delete an old file
        E = edit an old file
        F = file list from disk
        O = copy an old file
        Q = quit -- return to EXPLORE
        X = exit -- done with EXPLORE

====================================================
ENTER COMMAND =>
```

Figure 2.4   File Edit Menu

Perhaps you need to do something with one or more diskette data files. Then you type F on the main menu to choose the *file edit menu*. The file edit menu will appear on the screen, as shown in Figure 2.4. From this menu you can use C to *create* a new data file, D to *delete* a previously saved data file, E to *edit* an existing data file (that is, fix an incorrect entry, delete some data lines, insert new data lines, and save the changes), F to display the names of disk data files, or O to *copy* a previously saved data file. These tasks are collectively known as *file editing*. When finished, you may use Q to *quit* and return to the EXPLORE main menu, or X to *exit* the EXPLORE programs.

Although the most important function of the *file editor* is to let you create your own new data files, there are also many files already stored on the DATABASE-1 diskette. With the database you can use EXPLORE's data analysis programs without typing in your own data files from the keyboard. They can be modified by using the editor if you wish. Or you can add new data files to the database, using your own data sources.

When you type H from the main menu, you will get the *help menu,* shown in Figure 2.5. It offers various types of help, depending on your command.

```
=====================================================
HELP    MENU    COMMANDS

        D = disk file description
        F = file editing
        G = general help
        L = describe all programs
        P = describe one program
        Q = quit -- return to EXPLORE
        X = exit -- done with EXPLORE

=====================================================
        ENTER COMMAND ==>
```

Figure 2.5   Help Menu

Using this menu, you can tailor the help to your needs. The D command will print definitions of *data* stored on the DATABASE diskette. (If you are using your own data diskette, this command will not work.) The F command tells you about file editing and use. The G command gives general help. This consists of a series of screens of information. You may return to the help menu from any screen. The L command lists and describes all the EXPLORE programs. You will use the P command most often because you will want to describe a particular program. Finally, you may use Q to quit and return to the main EXPLORE menu or X to exit from the EXPLORE programs entirely.

An advance word of caution is in order: Do not rely on the help command to tell you everything! There is not enough diskette space to store everything you need to know. Even if there were, you would still need to study and plan carefully before attempting to use the computer. In other words, there is no substitute for careful advance reading in this manual or your statistics textbook. After a while, you will find it unnecessary to use the help command at all.

This concludes our overview of the menus. Now we will look more closely at the file editor. We begin by telling you how to input your own files, with examples to make things clearer. Then we discuss how the file editor can be used to examine files already on the DATABASE-1 diskette.

What Does a Data File Look Like?

The format of a data file is extremely simple. Each file is a column of numbers. Each data item occupies one line. At the end of each line you press the ENTER key, which is not printed but which causes the computer to look for the next line of data. The number of lines in the file equals the number of data points observed. For example, suppose you looked up the number of black city and county elected officials in each year during the period 1972–1980. You would have nine numbers. Figure 2.6 shows how your data file would look if you typed in those nine numbers. *Do not include the year*—only the data.

```
1108
1264
1602
1878
2274
2497
2595
2647
2832
```

Figure 2.6 Number of Black City Elected Officials, 1972–1980
Source: U.S. Bureau of the Census, *Statistical Abstract of the United States,* 1982–83 (Washington, D.C.: U.S. Government Printing Office, 1983), p. 488.

Your data file of N numbers will be stored on the diskette so that any EXPLORE program can read it. After reading the file, the program will temporarily store the data in the computer's memory until the program is finished. The actual diskette data file is not affected by being read. You can read the same file dozens of times, and it will still be there.

You can copy or delete files. You can even delete the DATABASE-1 files (either accidentally or deliberately). That would be a shame, since once they are gone, they are gone forever. Fortunately, the file editor warns you if you are about to delete a file and makes sure that you really want to. But do delete *unwanted* files periodically to avoid filling up your diskette.

■ How to Enter a Data File

Suppose you have been keeping track of your car's fuel economy (miles per gallon) and would like to do a statistical analysis of the last ten tanks of gas you used. Figure 2.7 shows how you would create a file called MPG for your ten data values. The file will not be displayed until you press a key.

Notice that the computer automatically capitalizes the file name you choose. All diskette files are assumed to have uppercase names. File names should be short (eight or fewer characters) and may contain almost any numeric or alphabetic characters. When you display your data set (called the *workfile*) on the screen, it will look like Figure 2.8. The file is also arranged horizontally, so more data can be displayed at once. The number of rows and columns depends on your data set. Unless your file is really huge, it will fit on one screen. Otherwise, it will be presented to you one screen at a time, offering you the option to stop with any screen to inspect your work so far.

A command line is given at the bottom of the data display. When you display the data you have entered so far, you are automatically placed in edit mode to permit you to *fix* incorrect entries (F), *delete* unwanted data entries (D), or *insert* additional data entries (I). When you are done editing, just type K. You are then given the option to save your file or to abandon it. After saving, you may either resume editing or return to the main menu to continue your analysis.

If you are working with a large file, you might want to stop to inspect it periodically and perhaps save it (using the K command) just in case anything unexpected should happen to your computer. In the present example our file MPG is so small that it is easy to do it in one step and to display it on one screen.

The data display is kept on the screen while you perform the editing, so you can see what is happening. Each time you make changes, the data display is erased, and the new (revised) version is displayed.

What do you do if you have made an error? You just fix it (using the F command). Suppose you discover that entry 9 should have been 24.2 instead of 22.4. You can correct this transposition easily, as shown in Figure 2.8.

```
===================================================

F I L E    E D I T    C O M M A N D    M E N U

        C = create a new file
        D = delete an old file
        E = edit an old file
        F = file list from disk
        O = copy an old file
        Q = quit -- return to EXPLORE
        X = exit -- done with EXPLORE


===================================================

ENTER COMMAND => c

NEW FILE TO BE CREATED ...

Name of new file? mpg

READY FOR INPUT ... Type one data item on each line
then press ENTER. When last item is typed, or if you
want to look at or fix your input so far, just press
ENTER (i.e. press the key with no input). You will
then be allowed to fix any errors you have made.

ITEM 1 ? 25.2
ITEM 2 ? 20.4
ITEM 3 ? 31.8
ITEM 4 ? 26.3
ITEM 5 ? 28.4
ITEM 6 ? 34.0
ITEM 7 ? 27.7
ITEM 8 ? 25.5
ITEM 9 ? 22.4
ITEM 10 ? 30.1
ITEM 11 ?

So far you've entered 10 items.
Maximum array size is 1000
Press any key to see data display
```

Figure 2.7   Creating a New Data File Named MPG

```
-----------------------------------------------------------------
                    L I S T    O F    D A T A
                        File = MPG
-----------------------------------------------------------------
  1: 25.2          4: 26.3          7: 27.7          10: 30.1
  2: 20.4          5: 28.4          8: 25.5
  3: 31.8          6: 34.0          9: 22.4
-----------------------------------------------------------------
COMMANDS:    F=fix      D= delete    I=insert    K=save/quit
COMMAND => f
Fix which item? 9
Current item 9   : 22.4
Revised item 9   ? 24.2
```

Figure 2.8   Fixing a Workfile Entry

```
----------------------------------------------------------------
                    L I S T   O F   D A T A
                        File = MPG
----------------------------------------------------------------
    1:  25.2        4:  26.3        7:  27.7       10:  30.1
    2:  20.4        5:  28.4        8:  25.5
    3:  31.8        6:  34.0        9:  24.2
----------------------------------------------------------------
COMMANDS:     F=fix     D= delete     I=insert     K=save/quit
COMMAND => i
Insert after which item (0 if at beginning of file) ? 9

READY TO INSERT ... Type one data item on each line
and then press ENTER.  When you are finished inserting
items, or want to look at your whole data set, just
press ENTER (i.e. press the key with no input).

Item 10 ? 28.8
Item 11 ? 27.3
Item 12 ?
```

Figure 2.9   Inserting Data Lines into Workfile

Now suppose you wish to add two more data lines, which you forgot to enter. Suppose these two tanks of gas (28.8 and 27.3 miles per gallon, respectively) happen to fall between the ninth and tenth data entries. You simply insert the new data (using the I command), as shown in Figure 2.9. The new data displayed shows the inserted lines.

Suppose you decide not to use the two additional lines after all. You can delete these same two lines (or any other lines) using the D command, as shown in Figure 2.10. If you want to delete only one item, just specify the same line number twice. If you delete anything, it will be lost forever (unless you have saved your file already), so be careful.

When you finish editing, you will want to save the workfile, as shown in Figure 2.11. You have the opportunity to change the file name before saving. Also, you will be warned if you use a filename that already exists (if you proceed anyway, the old file will be overwritten).

```
----------------------------------------------------------------
                    L I S T   O F   D A T A
                        File = MPG
----------------------------------------------------------------
    1:  25.2        4:  26.3        7:  27.7       10:  28.8
    2:  20.4        5:  28.4        8:  25.5       11:  27.3
    3:  31.8        6:  34.0        9:  24.2       12:  30.1
----------------------------------------------------------------
COMMANDS:     F=fix     D= delete     I=insert     K=save/quit
COMMAND => d
First item to delete ? 10
Last item to delete  ? 11
```

Figure 2.10   Deleting Lines from Workfile

```
------------------------------------------------------------
                   L I S T    O F    D A T A
                      File = MPG
------------------------------------------------------------
   1: 25.2         4: 26.3         7: 27.7        10: 30.1
   2: 20.4         5: 28.4         8: 25.5
   3: 31.8         6: 34.0         9: 24.2
------------------------------------------------------------
COMMANDS:    F=fix     D= delete    I=insert    K=save/quit
COMMAND => k


        -----------------------------------------------
          D A T A    F I L E    S A V E    M E N U

             D = done editing, ready to save file
             S = save file, continue editing
             Q = abandon file, quit editing

        -----------------------------------------------
        ENTER COMMAND => d

        Which drive for saving (A, B, C, or D) ? b
        Current working file name: MPG
        Change file name (Y or N)? n
        File B:MPG was saved.  Press any key to continue
```

Figure 2.11 Saving the Workfile

Notice that you have several *save* choices. In our example, file editing was completed, so the D command was used. But if this were a precautionary save and you wished to continue editing, you would use the S command. If the file becomes hopelessly botched, or if you decide not to use the file, you can abandon the workfile by choosing the Q command, in which case any previously saved version of the workfile will be intact.

You must have a disk ready to receive the file (usually in drive B), or an error message will be given. You will also get an error message if the disk has no room for your file. Notice the disk drive specification in the file name (B:MPG means that file MPG is stored on drive B). It is not necessary to specify a drive until you are ready to save the file.

You can save data files on the DATABASE-1 diskette, but be careful not to overwrite existing data files. You may want to keep an empty formatted diskette handy in case you fill up all the space on the one you are using. You can also get an error message if you have too many files on one disk, even though some space may exist. An error message also tells you if the disk drive door is open or if you forgot to insert the diskette.

Saving a file is the point when trouble is most likely to strike. The programs are designed to recognize some types of problems, but unforeseen situations do arise. At worst, an error might cause the program to terminate. You can start the program running again by typing START and then pressing the ENTER key.

After you have created a file, you can always call it back for further editing, as illustrated in Figure 2.12. The computer first lists your stored disk

```
===================================================
F I L E    E D I T    C O M M A N D    M E N U

            C = create a new file
            D = delete an old file
            E = edit an old file
            F = file list from disk
            O = copy an old file
            Q = quit -- return to EXPLORE
            X = exit -- done with EXPLORE


===================================================

ENTER COMMAND => e

LIST OF DISK DATA FILES WILL BE SHOWN ...
Which disk drive (A, B, C, or D)? b
-----------------------------------------------------------------
                  DATA FILES STORED ON DRIVE B
-----------------------------------------------------------------
PAYROLL       BUDGETS       PRICING       B-RATE       GNP-CAP
COUNTRY       DENSITY       MPG
-----------------------------------------------------------------
File to edit? mpg
```

Figure 2.12   Editing an Old Data File

```
====================================================
F I L E    E D I T    C O M M A N D    M E N U

            C = create a new file
            D = delete an old file
            E = edit an old file
            F = file list from disk
            O = copy an old file
            Q = quit -- return to EXPLORE
            X = exit -- done with EXPLORE


====================================================

ENTER COMMAND => o

FILE TO BE COPIED:

Copy from which drive (A, B, C, or D) ? b
Name of file to be copied ? mpg

FILE FOR SAVING COPY:

Copy to which drive (A, B, C, or D) ? b
Name of file to copy to ? mileage

CONFIRM COPY INSTRUCTIONS:

You want file B:MPG copied to file B:MILEAGE.
Type Y to confirm: y
Successful copy.
Press any key to return to main menu
```

Figure 2.13   Copying a Data File

```
==================================================
F I L E    E D I T    C O M M A N D    M E N U

          C = create a new file
          D = delete an old file
          E = edit an old file
          F = file list from disk
          O = copy an old file
          Q = quit -- return to EXPLORE
          X = exit -- done with EXPLORE


==================================================

ENTER COMMAND => d

LIST OF DISK DATA FILES WILL BE SHOWN ...
Which disk drive (A, B, C, or D)? b

--------------------------------------------------
             DATA FILES STORED ON DRIVE B
--------------------------------------------------
PAYROLL      BUDGETS      PRICING      B-RATE     GNP-CAP
COUNTRY      DENSITY      MPG          MILEAGE
--------------------------------------------------
File to delete ? mileage
Type Y to confirm: y
File B:MILEAGE has been deleted.
Press any key to return to main menu
```

Figure 2.14   Deleting a Data File

files, then asks you which one to edit. The file you name becomes the workfile. If you just want to look at a file, you can type E, as if you wanted to edit the file, and then use K to quit without saving the workfile.

Suppose you want to copy a file. Figure 2.13 shows how to do this. The original file MPG will be unchanged, and the new file MILEAGE will appear on the disk as a clone of MPG. Notice that the computer asks you to confirm the copy instructions, using the full names of the files. You may copy data files from one drive to another. Be forewarned, though, that this command will *not* copy anything except sequential data files. You cannot copy programs—in fact, programs will not even show up on your screen as saved files.

When you find that you have accumulated files you no longer need, it is easy to get rid of them. To delete a file, just follow the format shown in Figure 2.14. In this example we are deleting the file MILEAGE.

Data Conditioning

The EXPLORE programs, like many statistical programs, make the assumption that your data set will not contain excessive digits or numbers of unusual magnitude. Really huge numbers (such as 3,234,567) could cause loss of ac-

Table 2.1   Example of Data Conditioning

Year	U.S. Population	Conditioned Data File
1950	151,325,798	151.3
1960	179,323,175	179.3
1970	203,302,031	203.3
1980	226,545,805	226.5

Source: U.S. Bureau of the Census, *Statistical Abstract of the United States, 1984* (Washington, D.C.: U.S. Government Printing Office, 1984), p. 6.

curacy in some calculations or overflow of established print fields. It is better to express such numbers in millions, by shifting the decimal point as needed (3.234567). It may be appropriate to round off the data to fewer significant digits (3.235), since the accuracy is probably spurious to begin with and will have little effect on most calculations.

The opposite is true if the data are of extremely small magnitude (such as 0.000517). Such numbers should have their decimal point shifted as appropriate (0.517); later you will just have to remember that this was done. All that really is happening is that the units of measurement are being redefined for the data file. When you must shift the decimal point, be sure to make the same shift in all data points in the file, not just in the offending ones.

The example in Table 2.1 illustrates how raw numbers might be conditioned so as to reduce the chance of problems. The seeming loss of accuracy is probably unimportant, given that the U.S. Census has been shown to have rather large errors (on the order of several million people). That is to say, there really were not *exactly* 226,545,805 people in the United States in 1980. Many experts place the actual number several million higher. Given this uncertainty, rounding it off to 226.5 million (say, for the purpose of estimating a trend) will do no violence to trend calculations or to projections for the future. However, this depends on your data.

Data conditioning is not just a matter of decimal point shifting. Advanced data-conditioning techniques include log transformations, standardizing, and outlier trimming. You will learn more about these methods later on. For now, just remember that a decimal shift can often eliminate computing complications.

■ Case Study: World Demographic Profile

Table 2.2 shows data for twelve randomly chosen nations of the world: annual birth rate per 1000 persons, gross national product per capita (converted to

Table 2.2   Statistical Profile of Twelve Nations

Country	Birth Rate	GNP per Person	People per Square Mile
Afghanistan	32	170	66
Bangladesh	47	110	1665
Bolivia	47	890	12
Chad	44	220	10
Egypt	38	540	110
Guatemala	43	1150	174
Indonesia	43	460	192
Norway	13	14930	33
Spain	16	5600	200
Tanzania	47	230	54
U.S.A.	15	11590	64
Vietnam	42	150	434

Source: *Reader's Digest 1982 Almanac* (Pleasantville, N.Y.: Reader's Digest Association, Inc., 1982), pp. 467–475.

U.S. dollars at prevailing exchange rates), and population density (persons per square mile of land area).

The sample printout on pages 28–29 shows how to use the EXPLORE editor to create a file COUNTRY containing the names of these twelve nations. Using the same procedure, we could enter the three numerical data files using names such as the following:

B-RATE = birth rate per 1000 persons per year

GNP-CAP = gross national product per capita (dollars)

DENSITY = population density (persons per square mile)

Note that all variables are well-conditioned. Printouts for entering the three numerical files are not shown. The process would be identical with the creation of the file COUNTRY, except that we enter numbers instead of letters.

When you are entering a database, it is appropriate to include a labeling file such as COUNTRY to serve as an identifier when the files are listed using certain programs (such as ANALYZ, BIGRES, LAYOUT, MGRES, or SORTER). You might wish to glance ahead at the sample printouts for these programs. For example, if you look ahead to the sample printout of the program SORTER (page 237), you will see this same database used as an illustration. In SORTER we include COUNTRY as one of the files to be sorted on DENSITY.

```
===================================================
F I L E   E D I T   C O M M A N D   M E N U

        C = create a new file
        D = delete an old file
        E = edit an old file
        F = file list from disk
        O = copy an old file
        Q = quit -- return to EXPLORE
        X = exit -- done with EXPLORE


===================================================

ENTER COMMAND => c


NEW FILE TO BE CREATED ...

Name of new file? country

READY FOR INPUT ... Type one data item on each line
then press ENTER.  When last item is typed, or if you
want to look at or fix your input so far, just press
ENTER (i.e. press the key with no input). You will
then be allowed to fix any errors you have made.

ITEM 1 ? Afghanistan
ITEM 2 ? Bangladesh
ITEM 3 ? Bolivia
ITEM 4 ? Chad
ITEM 5 ? Egypt
ITEM 6 ? Guatemala
ITEM 7 ? Indonesia
ITEM 8 ? Norway
ITEM 9 ? Spain
ITEM 10 ? Tanzania
ITEM 11 ? U.S.A.
ITEM 12 ? Vietnam
ITEM 13 ?


So far you've entered 12 items.
Maximum array size is 1000
Press any key to see data display

-------------------------------------------------------------
                  L I S T   O F   D A T A
                     File = COUNTRY
-------------------------------------------------------------
  1: Afghanistan   4: Chad        7: Indonesia   10: Tanzania
  2: Bangladesh    5: Egypt       8: Norway      11: U.S.A.
  3: Bolivia       6: Guatemala   9: Spain       12: Vietnam
-------------------------------------------------------------
COMMANDS:    F=fix    D= delete    I=insert    K=save/quit
COMMAND => k

       -----------------------------------------------
         D A T A   F I L E   S A V E   M E N U

          D = done editing, ready to save file
          S = save file, continue editing
          Q = abandon file, quit editing

       -----------------------------------------------
```

```
ENTER COMMAND => d
Which drive for saving (A, B, C, or D) ? b
Current working file name: COUNTRY
Change file name (Y or N)? n
File B:COUNTRY was saved.
Press any key to continue
```

■ Exercises

2.1 To start things out, use the H command on the main EXPLORE menu to examine the help files. Familiarize yourself with what is available concerning the programs, files, and so on.

2.2 Next, call the EXPLORE file editor, using the main menu's F command. Satisfy yourself that you can then call back the main menu. Return to the file editor, and list the disk files on drive A (there should be none). Do the same for drive B.

2.3 To familiarize yourself with the creation of new files, enter a data file of your own, using the data set shown in Table 2.3. Do not enter the years—just the six numbers starting with 58.5 and ending with 47.4. Save your workfile on drive B in a diskette file called VOTING. Then copy it to a new file named NEWVOTE. Then delete the file NEWVOTE. This should leave you back where you started but with one extra file called VOTING. (Verify this by listing the files on the disk.)

Table 2.3   Percent of U.S. Voting Age Population Participating in Presidential Election

Year	Percent
1960	58.5
1964	57.8
1968	55.1
1972	50.7
1976	48.9
1980	47.4

Source: U.S. Bureau of the Census, *Statistical Abstract of the United States, 1984* (Washington, D.C.: U.S. Government Printing Office, 1984), p. 262.

2.4 Go back to the EXPLORE main menu and use the P command to run the program LAYOUT. Use LAYOUT to print your file VOTING. The intent is to give you a feeling for how the programs work. When finished, delete your file VOTING so that the diskette will have as much room as possible later on.

2.5 Familiarize yourself with the EXPLORE file editing system by using the file editor to edit one of the existing files in the state or time-series databases on the DATABASE-1 diskette. Try changing an entry, inserting some new lines, deleting a few lines, and whatever else you wish. Do *not* save the "edited" version, though (just abandon it by using Q on the file save menu). This exercise is a bit dangerous, but everything is fine as long as you do not save the edited file. The worst that can happen is that you will lose one data file, and that would not be the end of the world.

2.6 Check yourself on terminology by writing down from memory your own definitions of these terms: database, cross-sectional data, time-series data, file editing, workfile, menu.

3
ANALYZ/Analyzing Cross-Sectional Data

Introduction

One customary task of the statistician is to summarize the characteristics of an observed collection of n data points representing a cross-section of persons, firms, states, or any other units (but ordinarily *not* years or months—special procedures are needed to deal with time-series data). ANALYZ provides all the commonly used statistics of central tendency, dispersion, and shape (skewness and kurtosis) and presents them in a compact tabular format.

ANALYZ sorts the observed data and prints both the sorted original X array and the corresponding standardized Z values. Thus at a glance you can see how many standard deviations above or below the mean each X value lies, to check for symmetry or compare with a classic bell-shaped curve. An optional label file (names of states, firms, nations, or whatever) may be used to identify the cases. If any case identifier is longer than seven characters, it is truncated to fit in the space allocated by ANALYZ.

ANALYZ prints a histogram and frequency tabulation of your data grouped into classes. ANALYZ will do the histogram automatically if you wish or will offer you advice and let you set it up yourself. If you do the histogram yourself, ANALYZ will ask you to specify the starting point, interval width, and number of intervals. ANALYZ's recommendations for class size and number of classes are based on your data's skewness and whatever else is needed to achieve "nice" class limits for your data.

Although it is not apparent from the printout, ANALYZ's aesthetic algorithm for choosing class limits attempts to recommend choices like those a human would make, based on our preference for "round" class limits and "nice" class intervals. You might find it interesting to try to improve on the computer's recommendations. Sometimes this is quite easy—the algorithm has its blind spots and can be fooled by certain data sets. Many times, though, you may like the computer's choices better than any you can come up with.

Appendixes F and G can assist you in analyzing skewness and kurtosis (the shape of the sample histogram). These two appendixes contain ranges for the skewness and kurtosis coefficients of samples drawn from normal bell-shaped populations.

■ Case Study 3.1: Price-Earnings Ratios for Standard and Poor's 500 Stocks

In the stock market, investors pay close attention to a company's price-earnings (PE) ratio, defined as current stock price divided by estimated new year earnings (or last twelve months' earnings if no estimate is available). This allows a comparison between the market's valuation of the stock and the actual earnings prospects. What do these ratios look like? What is their range? What values are typical?

Recent price-earnings ratios were chosen at random from the population of all companies listed in *Standard and Poor's Stock Guide.* The sample was chosen from a monthly stock guide by picking the first company at the top of each page divisible by 10 (page 10, page 20, page 30, and so on). Since the guide contained 242 pages, the resulting sample contains twenty-four companies. Two companies showed negative earnings, so the next company on the page was chosen instead. The principal business of each company was noted. The data are summarized in Table 3.1.

These twenty-four price-earnings ratios were entered into a data file named PE. Using program ANALYZ, we generate the sample printout.

```
                              ANALYZ
        This program will compute and display the most commonly
needed descriptive statistics for one variable.  Your data
must already be stored in a sequential data file as a single
column of N observations.  Maximum is N = 1000.  A histogram
and sorted array with standardized values are also provided.
If desired, labels may be read from a second file of N items
to identify cases (only first 7 characters will be printed).

Name of file to analyze? pe
Do you want to use a label file (y/n)? y
Name of label file? company
Computing ...


GENERAL FACTS:                 DISPERSION:

    File = PE                      Variance = 35.08515 (33.62327)
    Observations =   24            St. Dev. = 5.923272 (5.798557)
    Minimum = 5                    Coef. of Var. = 53.64472 (52.51523)
    Maximum = 31                   Avg. Dev. About Mean = 3.722223
    Range= 26                      Avg. Dev. About Median = 3.541667


CENTRAL TENDENCY:              SKEWNESS AND KURTOIS:

    Mean = 11.04167                2nd Moment = 33.62326
    Median = 9.5                   3rd Moment = 442.8388
    Geom. Mean = 10.04011          4th Moment = 8831.464
    1st Quartile = 7               Skewness =   2.271 (Pearson Beta 1)
    2nd Quartile = 9.5             Kurtosis =   7.812 (Pearson Beta 2)
    3rd Quartile = 12.5            1 Outlier(s) Detected

            Note: Standard deviation using N instead
                  of N-1 shown in parentheses.
```

Table 3.1   Price-Earnings Ratios of Selected Major U.S. Companies

Company	Principal Business	PE Ratio
Aetna Life & Casualty Co.	multiplan insurance	9
Ameritech Corp.	telephone service	7
Bank of Boston	bank holding	5
Brunswick Corp.	marine, recreation products	10
Chesebrough-Ponds	health, beauty, food, apparel	11
ConAgra Inc.	bakery flour, feed, poultry	13
Denny's Inc.	restaurants, donut shops	12
Electronic Data Systems	business information systems	26
First Mississippi	chemicals, fertilizer, oil, gas	31
Gillette Co.	shaving, personal care, pens	10
Hilton Hotels	hotels, casinos	13
InterNorth Inc.	international oil exploration	9
Leaseway Transport	motor vehicle transportation	11
Masonite Corporation	hardboard manufacturing	13
Morgan J.P.	bank holding	7
NYNEX Corp.	telephone service	7
Penney J.C.	department stores, mail order	8
Public Sv. El. & Gas	utility: electricity and gas	7
Ryder Systems	vehicle leasing, rental	12
Sonat Inc.	gas pipeline, oil and gas drill	7
Taft Broadcasting	TV/radio, TV films, amusement packages	13
Travelers Corp.	multiline insurance	8
V F Corp.	intimate apparel, leisure	7
Woolworth F.W.	variety and discount stores	9

Source: *Standard & Poor's Stock Guide* (New York: Standard & Poor's Corporation, January 1984). Data are provided courtesy of Kidder, Peabody Inc.

The first screen contains all the common statistics of central tendency (where is the middle or typical value of the data?), dispersion (how spread out are the data?), and skewness and kurtosis (what is the shape of the distribution?). These statistics tell the statistician a great deal.

From ANALYZ's sorted array of data points we look for the most frequently occurring values. We find that the *mode* is 7, which occurs six times. However, there are also modes at 9 and 13, suggesting that the modal PE ratio may not give a reliable indication of central tendency. This is an inherent flaw of the mode: It is not necessarily unique, especially in small samples. On the other hand, looking ahead at the highest bar on the printout's histogram, the modal *class* in the frequency classification is probably a good measure of central tendency (the range 5 to 10 on the histogram includes twelve companies).

We find that the range of PE ratios is from 5 (Bank of Boston) to 31 (First Mississippi). ANALYZ's *sample mean* is 11.0, while the *sample median* is 9.5, suggesting that the distribution is skewed right (the mean is "pulled up" above the median by one or more large values). Since the median is unaffected by *outliers* (unusually high or low data values), it has appeal as a measure of central tendency in a situation such as ours, in which a few unusually high values are present. Using the definition of the median, we can say that about half the firms had PE ratios below 9.5 (and the other half were above 9.5).

The *geometric mean* is 10.0, which lies between the arithmetic mean and the median. Ordinarily, we use the geometric mean (defined as the nth root of the product of the data values) to average ratios or percents. Although business statisticians do not often refer to it, the geometric mean has certain advantages. For example, the geometric mean is less affected by high positive outliers (extremely large data values). Since we suspect that outliers are present, the geometric mean does give a useful reference point for our data.

The *quartile points* offer insight into both central tendency and dispersion, because they divide the sorted data points into groups that are easily interpreted (see Figure 3.1). The second quartile, of course, is simply the median. The first quartile is the median of the lower half of the sample, and the third quartile is the median of the upper half. For our data, ANALYZ shows that the first quartile point is 7 and the third quartile point is 12.5, so it is approximately correct to say that the middle 50 percent of companies have PE ratios between 7 and 13. Because our sample size is not very large, and because of multiple occurrences of the same data values, the quartile points are only approximate in our sample.

The *coefficient of variation* is 53.6. Since the coefficient of variation is simply the ratio of the standard deviation to the mean (expressed in percent), we can say in this case that the standard deviation is more than half the size of the mean. This suggests a high degree of dispersion.

This high dispersion is not unexpected, since two companies with unusually high PE ratios (First Mississippi at 31 and Electronic Data Systems at 26) catch our attention even before we look at the computer results from ANALYZ. It is approximately true to say that a "two-sigma" interval about the mean (that is, mean plus or minus two standard deviations) ordinarily will enclose about 95 percent of the sample data. But our standard deviation (5.9)

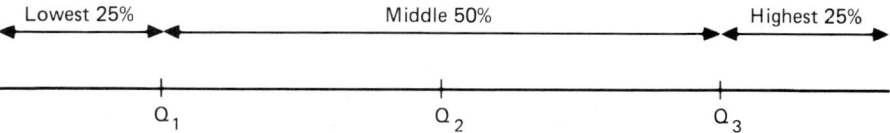

Figure 3.1 Quartile Points

is so large that a two-sigma interval would range from almost zero up to 20. Such an interval would be almost meaninglessly wide, in terms of predicting where a company's PE might be expected to fall.

ANALYZ's calculated *skewness coefficient* (2.271) reveals strong right-skewness. A skewness coefficient near zero would indicate near-symmetry. As discussed in Appendix F, if the population being sampled were bell-shaped, we would expect this statistic to lie between $-.711$ and $+.711$ most of the time, allowing for sample variation. Since our skewness coefficient lies well outside that range, the sample skewness is significant. The main cause is probably the outlier First Mississippi, which is more than three standard deviations from the sample mean.

ANALYZ's *kurtosis coefficient* (7.812) suggests a strongly *leptokurtic* shape (more peaked than a normal bell-shaped curve). A perfect bell-shaped (or *mesokurtic*) normal population would have a kurtosis coefficient of exactly 3, while flatter than normal (or *platykurtic*) populations have kurtosis coefficients below 3. Although Appendix G does not have a kurtosis coefficient range for samples as small as n = 24, our kurtosis statistic appears well beyond the upper end of the range even for larger samples.

ANALYZ's formulas for the skewness and kurtosis coefficients are based on the second, third, and fourth moments about the sample mean:

$$\text{Skewness coefficient} = M_3/M_2^{3/2}$$

$$\text{Kurtosis coefficient} = M_4/M_2^2$$

where the jth moment about the sample mean is defined as

$$M_j = \frac{1}{n} \sum_{i=1}^{n} (X_i - \overline{X})^j \quad \text{for } j = 2, 3, 4.$$

Although ANALYZ handles all the calculations, you may want these formulas for reference. Appendixes F and G have also been prepared to assist you in interpreting the results shown on the ANALYZ printout.

SORTED AND STANDARDIZED PE VALUES

Rank	Case	X	Z	Rank	Case	X	Z
1	Bk Bost	5	-1.020	13	Brnswk	10	-0.176
2	NYNEX	7	-0.682	14	Gillett	10	-0.176
3	Amertch	7	-0.682	15	Chsb-Pn	11	-0.007
4	VF Corp	7	-0.682	16	Leasewy	11	-0.007
5	PSE&G	7	-0.682	17	Denny's	12	0.162
6	Sonat	7	-0.682	18	Ryder	12	0.162
7	MorganJ	7	-0.682	19	Masonit	13	0.331
8	Travlrs	8	-0.514	20	Taft	13	0.331
9	Penney	8	-0.514	21	Hilton	13	0.331
10	Aetna	9	-0.345	22	ConAgra	13	0.331
11	IntNrth	9	-0.345	23	EDS	26	2.525
12	Woolwth	9	-0.345	24	Fir Mis	31	3.369

[Note: Z transform uses sample standard deviation]

Table 3.2   Frequency of Occurrence of X Values

Range of PE Values	Actual Number	Actual Percent	Normal Percent
Within 1 S.D. of Mean	21 out of 24	88	68
Within 2 S.D. of Mean	22 out of 24	92	95
Within 3 S.D. of Mean	23 out of 24	96	99

Now, looking at ANALYZ's sorted array, we see a gap between 13 and 26. Moreover, the standardized z values reveal a distinct lack of symmetry: Eight companies are above the mean, while sixteen are below the mean. The median would give a much better indication of central tendency, it seems. The two unusually high z values ($z = 2.525$ and $z = 3.369$) stand out clearly. In a bell-shaped normal distribution, it is rare to find z values that exceed 2.0 and extremely rare to encounter z values greater than 3.0. You would want to look closely at these two companies (EDS and First Mississippi) to find out if there was something unusual in their recent earnings records, management changes, or future prospects that might explain their peculiar PE ratios. (Hint: EDS belonged to H. Ross Perot at the time.)

Outliers (usually defined as z values greater than 3) are troublesome. There is a temptation to discard these extreme points because they are different from the other data points. Yet outliers often contain information without which the analysis is incomplete. Statisticians have developed various rules for handling outliers (there are entire books on the subject), but common sense must suffice for now. Since some firms really do have super-high PE ratios, we will simply accept this as part of our findings.

In addition to its asymmetry, the sample is quite different from a normal bell-shaped model in terms of points lying within various distances of the mean. We can tabulate the z values from ANALYZ's sorted array as shown in Table 3.2. Clearly, there are too many data points close to the mean. Our sample's histogram will not look very much like a normal bell-shaped curve.

As expected, ANALYZ's histogram is not bell-shaped at all. The points are clustered at the low end of the scale, while two unusually high values uphold the mean at the top of the scale. A revised histogram might show more detail at the lower end, but only if the two upper points were discarded.

```
HISTOGRAM SET-UP

To construct the histogram, shall the computer:

    1. make all decisions automatically
    2. offer advice, but leave decisions to you
    3. leave decisions to you entirely
 ?1
```

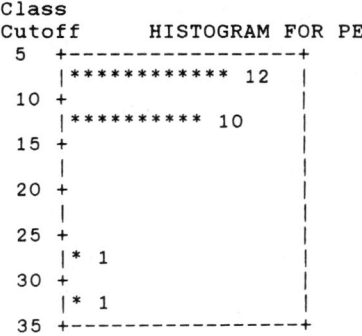

```
Class
Cutoff       HISTOGRAM FOR PE
  5 +-----------------+
    |************ 12  |
 10 +                 |
    |********** 10    |
 15 +                 |
    |                 |
 20 +                 |
    |                 |
 25 +                 |
    |* 1              |
 30 +                 |
    |* 1              |
 35 +-----------------+
```

```
Want to try the histogram again (y or n)? n

Where now:  E=EXPLORE menu   R=Run ANALYZ again   X=exit ? e
```

In choosing the number of histogram classes, ANALYZ starts with *Sturges's Rule,* which says that the ideal number of classes should be approximately:

$$\text{Ideal number of histogram classes} = 1 + \log_2(n).$$

Table 3.3 shows the approximate number of classes that Sturges's Rule would suggest for samples of various sizes. More classes should be used if the data are skewed. Like any rule of thumb, Sturges's Rule is only a guideline. We would feel free to depart from it somewhat in order to obtain "nicer" class intervals.

ANALYZ used six classes for our PE data, which is what Sturges's Rule would suggest. Since our sample is skewed to the right, perhaps more classes should have been used; however, the class limits chosen by ANALYZ appear quite reasonable, so we are satisfied.

On the basis of our random sample it might be said that a PE ratio of around 10 is typical for Standard and Poor's 500 companies. Companies with

Table 3.3   Ideal Number of Classes Using Sturges's Rule

Sample Size	Number of Classes
8	4
16	5
32	6
64	7
128	8
256	9
512	10

PE ratios below 7 are in the bottom quartile, while those at 13 or above are in the upper quartile. Outliers cause the distribution to be right-skewed and leptokurtic. A larger sample might reveal additional details. Also, a sample taken later in the year might be based on more reliable earnings figures.

■ Exercises

3.1 For your analysis, choose one or more of the state data files saved on your DATABASE-1 diskette. Your sample size will then be n = 50. Use the computer help menu and/or Appendix H to assist you in making your choice.

3.2 Run ANALYZ using your chosen data file, labeling cases using the file STATE. Experiment with different histograms if necessary to obtain a satisfactory idea of the shape of the distribution. Obtain printed copy of the results (assuming a printer is available) by pressing the PrtSc key each time a new screen is displayed. Optionally, use SAVON and SAVOFF. If you have no printing capability, plan to rerun the programs as needed to obtain information from the screen, and take notes so that you can complete Exercises 3.3–3.5.

3.3 Write a short nonstatistical description of your data. Discuss the units of measurement. Do you regard the data as a sample or as a population? Who probably collected the data, and why? What biases, errors, or omissions might have occurred during sample design or data gathering? What time period is covered? What other concerns could be raised about the data themselves?

3.4 Using the case study as a model, write a succinct statistical analysis of central tendency, dispersion, skewness, and kurtosis. In nontechnical language, what are your main conclusions about your data set's shape? Are there outliers? Could the data be from a normal population? Why, or why not? Use Appendixes F and G to check skewness and kurtosis for samples of size 50. What other conclusions can you reach from a thorough "eyeball" check of the data?

3.5 What does the histogram tell you? Is it consistent with your expectations, based on the skewness and kurtosis coefficients calculated by the computer?

3.6 Run BOXPLOT, using the same data. Describe the result in a short written synopsis. How does it compare with ANALYZ in terms of usefulness in succinctly describing a sample?

3.7 Assume that you have been asked to prepare a concise (one-page) report of your findings for a busy executive who has never seen the raw data and whose statistical background is not too great. Try to cover the pertinent points without getting bogged down in detail. When finished, try to anticipate this executive's reaction to your report (that is, be critical of your own report). What else could you add if you had plenty of time?

3.8   What would a busy decision maker gain from your report? What would he or she lose by relying solely on your report, without having the raw data? What instructions would *you* give to someone you were asking to prepare such a report? What contribution did the computer make to your report?

3.9   Identify the three states at the top of the distribution and the three states at the bottom. Suggest some reasons why these particular states occupy these relative positions. Where does your own state lie? Why?

3.10 Use the EXPLORE file editor to create two data files named CONC72 and CONC77. Data file CONC72 should contain the 1972 concentration ratios for the thirty-one industries shown in Table 3.4. Data file CONC77 should contain the concentration ratios for these same thirty-one industries in 1977. Use ANALYZ to obtain summary statistics for each file. Did the degree of concentration change noticeably between these two points in time? Fully discuss central tendency, dispersion, and shape of the data for both years, following the format of Exercises 3.3–3.5. Write a clear, concise, nontechnical summary of your findings, intended for a busy decision maker.

Overall, which industries are most concentrated in the top four firms? Which are least concentrated? Can you suggest reasons why certain types of industries are more concentrated than others? In one column, list the industries that experienced an increase in concentration. In a second column, list those that experienced a decrease. How many industries increased in concentration? Which ones increased? How many decreased? Which ones decreased? Why?

If we also considered foreign competitors, how would our image of concentration change for these industries? Discuss fully. How important are these nonstatistical questions?

Optional Exercises for Ambitious Learners

3.11 Identify two state data files from the DATABASE-1 diskette that you think may be related (for example, UNEM and DIVORCE). Run SORTER to list these two state data files side by side, sorting the second file on the first and obtaining printed copy if possible.

Use the first variable to divide all the states into two groups (for example, High Unemployment States and Low Unemployment States). Base these categories on a reasonable cutoff point—for example, High Unemployment States might be those with at least 7 percent unemployment, and Low Unemployment States would be the rest. Obviously, a cutoff point close to the median would be preferred, but a round approximation may make things easier.

Table 3.4   Percent of Shipments, in Value, by the Four Largest Firms
in Selected U.S. Industries, 1972 and 1977

Industry	1972	1977
Blast furnaces and steel mills	45	45
Motor vehicles and car bodies	93	93
Motor vehicle parts	61	62
Radio and TV communication equipment	19	20
Aircraft	66	59
Petroleum refining	31	30
Newspapers	17	19
Pharmaceuticals	26	24
Bread, cake, and related products	29	33
Photographic equipment and supplies	74	72
Fluid milk	18	18
Meat-packing plants	22	19
Construction machinery	43	47
Periodicals	26	22
Electronic computing equipment	51	44
Plastic materials and resins	27	22
Cigarettes	84	81
Bottled and canned soft drinks	14	15
Gray iron foundries	34	34
Sawmills	18	17
Papermills	24	23
Refrigeration and heating equipment	40	41
Soap and other detergents	62	59
Toilet preparations	38	40
Automotive stampings	69	65
Organic fibers, noncellulosic	74	78
Commercial printing, letterpress	14	14
Commercial printing, lithographic	4	6
Farm machinery and equipment	47	46
Tires and inner tubes	73	70
Guided missiles and space vehicles	62	64

Source: *U.S. Bureau of the Census, Statistical Abstract of the United States, 1977* (Washington, D.C.: U.S. Government Printing Office, 1977); and U.S. Bureau of the Census, *Statistical Abstract of the United States, 1982* (Washington, D.C.: U.S. Government Printing Office, 1982).

Use the EXPLORE file editor to type in two new data files containing the observed values of the second variable for each of these two groups. For example, if your second variable is DIVORCE, you could put divorce rates for the twenty-three High Unemployment States into a file called DIVORCE1, while divorce rates for the twenty-seven Low Unemployment States might go into a file called DIVORCE2. Alternately, you could do the same thing with program SPLIT.

Now run ANALYZ (and BOXPLOT if desired) for each of your two new files. See whether these two subsamples differ in central tendency, dispersion, or shape. If so, the result is presumably due to the effect of your classification. For example, are divorce rates higher in the states with high unemployment rates? Write a complete analysis of your findings, and suggest possible explanations for any noticeable differences between the two subsamples.

3.12 From a home almanac, choose any sample of data for n counties in your state (n depends on your state), n cities in the United States (n may be chosen as you wish), or n countries of the world (n depends on what is available). Use the EXPLORE file editor to enter your data into a diskette data file, choosing any file name except ones that might conflict with file names already on the diskette. Repeat Exercises 3.2–3.8 using these data.

3.13 Take a survey, asking some kind of numerical question of n of your friends or classmates (such as age, height, weight, number of children desired, number of traffic tickets received in past twelve months, number of days spent outside of the United States in last twelve months). Sample size depends only on your energy. Use the EXPLORE file editor to enter your sample results into a diskette data file, choosing any file name except ones that might conflict with those already on the diskette. Repeat Exercises 3.2–3.8.

References

1. Chambers, John M., *Computational Methods for Data Analysis* (New York: John Wiley and Sons, Inc., 1977), pp. 222–223.

2. Doane, David P., "Aesthetic Frequency Classifications," *The American Statistician,* Vol. 30, No. 4, 1976, pp. 181–183.

3. Ehrenberg, A. S. C., *Data Reduction: Analyzing and Interpreting Statistical Data* (London: John Wiley and Sons, Inc., 1975).

4. Sturges, Herbert A., "The Choice of a Class Interval," *Journal of the American Statistical Association,* March 1926, p. 65.

4
ANOVA/Analysis of Variance

Introduction

ANOVA (an abbreviation for *analysis of variance*) allows you to check for significant differences of means in observed sample data. The differences in means are assumed to arise because of certain causal factors, expressed by classifying the data into several categories (such as Group 1, Group 2, Group 3). Categories do not necessarily imply numerically scaled measurements and often are just qualitative labels (such as Brand A, Brand B, Brand C). Therefore ANOVA is useful in a variety of experimental situations in which inter-group variation is thought to exist.

Variation in observed data is assumed to be due to *one factor* (column effect) or *two factors* (row and column effects). The two models that the program ANOVA can analyze are:

1. One-way ANOVA (data classified into c columns, with n observations in each column);
2. Two-way ANOVA (data classified into c columns and r rows, with n observations in each row-column cell).

Clearly, the two-way ANOVA model is more complicated because of the variety of possible effects that might be studied. All ANOVA hypothesis tests utilize the F statistic, which shows the ratio of mean sums of squared deviations attributable to various possible sources. Appendix E contains critical values of the F statistic.

■ Case Study 4.1: One-Way ANOVA of Days Worked

The XYZ Corporation is interested in possible differences in absenteeism among salaried employees in various departments. A trial survey of twenty randomly chosen employees reveals the data shown in Table 4.1.

Because of the casual sampling methodology in this preliminary survey, the sample sizes are not equal. The sample data may be summarized as shown in Table 4.2.

Table 4.1   Days Worked Last Year by Twenty Randomly Chosen Employees

Budgets Department	Payroll Department		Pricing Department
278	205	217	240
260	270	266	258
265	220	239	233
245	240	240	256
	255	228	223
			242

The one-way ANOVA model is:

$$X = \mu + \Theta_i + \text{random error}$$

where

X = days worked by a particular employee;
μ = overall mean days worked by all employees;
Θ_i = effect of employee's department ($i = 1,2,3$).

The model implies that an employee's work days differ from the overall mean only because he or she belongs to a certain department. If all departments are the same, we would expect their effects to be zero; that is, we would expect the *null hypothesis* to be true, as stated in the following pair of hypotheses:

H_0: $\Theta_1 = \Theta_2 = \Theta_3 = 0$.
H_1: Not all the means are equal.

These two hypotheses may be stated equivalently, using the terminology of group means:

H_0: $\mu_{\text{Budgets}} = \mu_{\text{Payroll}} = \mu_{\text{Pricing}}$.
H_1: Not all the means are equal.

If the sample evidence strongly contradicts the null hypothesis H_0, we will reject the null hypothesis. Otherwise, we will provisionally accept the null hypothesis unless additional sample data become available.

Table 4.2   Mean Days Worked in Three Departments

Department	Mean Days Worked	Sample Size
Budgets	$\overline{X}_1 = 262$ days	$n_1 = 4$ employees
Payroll	$\overline{X}_2 = 238$ days	$n_2 = 10$ employees
Pricing	$\overline{X}_3 = 242$ days	$n_3 = 6$ employees
Overall	$\overline{\overline{X}} = 244$ days	$n = 20$ employees

ANOVA carries out the calculations illustrated in the sample printout (see page 45). The computer breaks down the total variation about the mean into its two component sources: variation *between* columns (attributable to differences between departments) and variation *within* columns (attributable to variation among individual employees in each department). The breakdown of total variation can be expressed mathematically as

$$SST = SSB + SSW$$

where

SST = total sum of squares [d.f. = n − 1];
SSB = between columns sum of squares [d.f. = c − 1];
SSW = within columns sum of squares [d.f. = n − c].

The F test statistic is then calculated by the computer as:

$$F = [SSB/(c − 1)]/[SSW/(n − c)].$$

Since the computer handles the calculations, you can concentrate your attention on the meaning of this formula: The F ratio measures the degree of variation *between* departments in comparison to variation *within* departments. The larger this F ratio, the more likely we are to reject H_0.

The observed differences in means are not convincing, since ANOVA yields a small test statistic: F = 2.589 with d.f. = (2,17). Since d.f. = (2,17) is not available in Appendix E, we actually look up d.f. = (2,15), using the next lower degrees of freedom (a conservative decision). At the .05 level of significance, the critical value of F is 3.68 (from Appendix E). Since F < 3.68, we cannot reject the null hypothesis. The decision rule is illustrated in Figure 4.1.

There are two reasons why we cannot reject the null hypothesis. First, the sample sizes are rather small (particularly in Budgets, which has only four employees sampled). Second, considerable variation exists within departments (note especially the high and low extremes in Payroll, which are greater than either of the other two departments). These facts weaken the inference. Perhaps a larger sample should be taken if this suggestive sample has aroused the curiosity of management.

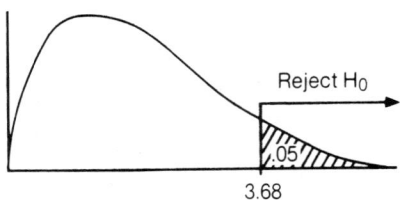

Figure 4.1   F Test for One-Way ANOVA

ANOVA

This program performs analysis of variance (ANOVA) to study relationships among columns and/or rows of a grouping of factors in an experiment. This program will handle two common experimental designs:

1. One-Way ANOVA   (column effects only)
2. Two-way ANOVA   (both row and column effects)

More details will be given once you have chosen your option. Which option do you want (1 or 2)? 1

ONE-WAY ANALYSIS OF VARIANCE

This program will read up to 8 saved data files and will carry out a test for equality of means. The hypotheses to be tested are:

HO: Mean(1) = Mean(2) = ... = Mean(c)
H1: Not all the means are equal

where c is the number of files being tested. Computer will print the means, standard deviations, between and within sums-of-squares, and F statistic. Files may be of unequal length (up to 500 cases per file).

FILES TO BE COMPARED

How many files? 3
Name of file # 1 ? budget
Name of file # 2 ? payroll
Name of file # 3 ? pricing

TABLE OF MEANS AND STANDARD DEVIATIONS
==
File	Mean	S.D.	Sample Size
BUDGET	262	13.63818	4
PAYROLL	238	21.23938	10
PRICING	242	13.40149	6
--------	-----	----------	----
Overall:	244	19.46116	20

A N O V A   T A B L E

==
Source	Sum of Squares	D.F.	Variance
Between:	1680	2	840
Within:	5516	17	324.4706
--------	------	----	----------
Total:	7196	19	378.7369

Computed F = 2.589 with d.f. = 2 and 17

Where now: E=EXPLORE menu   R=Run ANOVA again   X=exit ? e

■ Case Study 4.2: Two-Way ANOVA of Restaurant Ratings

Big Burger, a large fast-food chain, hires an outside consulting firm to carry out an objective rating of three Big Burger retail outlets in a certain metropolitan area. The consulting firm sets up a scoring system based on criteria such as cleanliness, friendliness, promptness, food appearance, food freshness, and food flavor. Five raters are assigned to rate each of the three retail facilities twice, making a total of thirty observations altogether. These thirty ratings (shown in Table 4.3) comprise a table with three columns (c = 3) and five rows (r = 5). Each cell has two observations (n = 2), since each restaurant is rated twice. This is a two-way ANOVA problem, since it involves both row and column effects.

It is apparent that some raters are tougher than others. It also looks as if some raters are inconsistent compared to others. The main hypotheses of interest concern column effects (are the three facilities rated differently?), row effects (do the five raters disagree?), and interaction effects (does some combination of facility/rater affect variation in ratings?). The implied two-way ANOVA model is

$$X = \mu + \Theta_i + \phi_j + \gamma_{ij} + \text{random error}$$

where

X = rating of an individual restaurant

μ = common mean rating for all restaurants

Θ_i = effect attributed to restaurant (i = 1,2, . . . , c)

ϕ_j = effect attributed to rater (j = 1,2, . . . , r)

γ_{ij} = effect of interaction between restaurant and rater (row i, column j).

Table 4.3   Ratings of Three Fast-Food Facilities

Rater	Facility 1	Facility 2	Facility 3
1	80	80	70
	82	78	72
2	86	77	70
	82	81	60
3	90	81	86
	94	81	84
4	86	74	70
	88	84	84
5	74	72	63
	78	72	61

The interaction effect γ_{ij} is absent if there is no replication (that is, if n = 1). The null hypotheses would be that these effects are zero (that Θ_i = 0 and/or ϕ_j = 0 and/or γ_{ij} = 0). If all effects are absent, the model simply collapses to a common mean rating:

$$X = \mu + \text{random error.}$$

But if the calculated F ratios are large enough, we can conclude that significant effects do exist.

We can rearrange the given data as shown in Table 4.4 by adding each rater's two ratings of each restaurant, giving a table of sums (see the ANOVA printout on pages 49–50). From these sums we can calculate means for each rater (row) and each restaurant (column). This table suggests that facility 1 is highly rated, while facility 3 is not so highly rated. It appears that rater 5 is hard to please, while rater 3 is inclined to be generous.

The computer breaks down the total variation about the mean into its components, using the following relationship:

$$SST = SSC + SSR + SSI + SSE$$

where

SST = total sum of squares [d.f. = rcn − 1]
SSC = between columns sum of squares [d.f. = c − 1]
SSR = between rows sum of squares [d.f. = r − 1]
SSI = interaction sum of squares [d.f. = (c − 1)(r − 1)]
SSE = error sum of squares [d.f. = rc(n − 1)].

Table 4.4   Table of Sums for Each Cell

Rater	Facility 1	Facility 2	Facility 3	Row Total	Row Mean
1	162	158	142	462	77
2	168	158	130	456	76
3	184	162	170	516	86
4	174	158	154	486	81
5	152	144	124	420	70
Column Total:	840	780	720	2340	
Column Mean:	84	78	72		78

The computer then calculates several F ratios, using these formulas:

Between columns: $F = [SSC/(c - 1)]/[SSE/(rcn - rc)]$

Between rows: $F = [SSR/(r - 1)]/[SSE/(rcn - rc)]$

Interaction: $F = [SSI/c - 1)(r - 1)]/SSE/(rcn - rc)]$.

If $n = 1$ (as in a randomized block design), these formulas are automatically modified by ANOVA so that d.f. $= (r - 1)(c - 1)$ for the denominator SSE, and the interaction term is ignored. The calculations are entirely carried out by the computer, so you can focus your attention on the *meaning* of these test statistics: The larger the F ratio, the more likely it is that we will reject the null hypothesis (which states that there is no effect). In other words, a large F ratio shows that the sample is *not* in agreement with the null hypothesis, suggesting that a significant effect does exist.

Using right-tail F tests at the .05 level of significance and degrees of freedom from the ANOVA printout, Appendix E yields the decision rules shown in Table 4.5.

From the ANOVA printout the calculated F ratios are

Between columns: $F = 22.31$

Between rows: $F = 13.20$

Interaction: $F = 1.74$.

Thus we conclude that there are significant row and column effects but no interaction effects. Overall, the classification scheme does help explain the observed ratings. Specifically,

1. Ratings of the three Big Burger facilities *do* differ significantly (column effect exists).
2. Raters *do* disagree among themselves (row effect exists).
3. Combination effects are *not* useful predictors of ratings (interaction effect does not exist).

If there is no replication ($n = 1$ observation per cell), the two-way ANOVA is computationally equivalent to a *randomized block design*. The interaction effect is suppressed by ANOVA in this case, and appropriate formulas are used.

Table 4.5   Decision Rules for Two-Way ANOVA

Effect Due To:	Decision Rule:	Degrees of Freedom
Column (Facility)	significant if F > 3.68	d.f. = 2,15
Row (Rater)	significant if F > 3.06	d.f. = 4,15
Interaction	significant if F > 2.64	d.f. = 8,15

ANOVA

   This program performs analysis of variance (ANOVA) to study relationships among columns and/or rows of a grouping of factors in an experiment. This program will handle two common experimental designs:

   1. One-Way ANOVA   (column effects only)
   2. Two-way ANOVA   (both row and column effects)

More details will be given once you have chosen your option. Which option do you want (1 or 2)? 2

TWO-WAY ANALYSIS OF VARIANCE

   This program will carry out a test of the two-way ANOVA model with up to eight rows and five columns, with n observations per cell. This is to test the model:

$X(i,j,k) = Mu + R(i) + C(j) + W(i,j) + E(i,j,k)$ where:

   $X(i,j,k)$ = observed value in row i and column j
                  $(k=1,2,...,n)$
       Mu = the intercept
     $R(i)$ = the effect of row i $(i=1,2,...,r)$
     $C(j)$ = the effect of column j $(j=1,2,...,c)$
   $W(i,j)$ = effect of interaction between
              row i and column j (omitted if n=1)
   $E(i,j,k)$ = random error for observation i,j,k

HYPOTHESES AND DATA FORMAT

Three sets of hypotheses are tested:

   Row Means: HO: R(i)=0 for all i
                 H1: Not all R(i) are zero

Column Means: HO: C(j)=0 for all j
                 H1: Not all C(j) are zero

 Interaction: HO: W(i,j)=0 for all i,j
 (if needed)   H1: Not all W(i,j) are zero

TWO-WAY ANOVA TABLE SET-UP AND FILE FORMAT

   Observations are read from c files (one for each column). Each file contains one column of data for the table, consisting of consecutive groups of n observations (each file therefore contains r*n observations). If there is only one observation per cell (non-replicated experiment, equivalent to randomized block) each file is a column of r observations, one for each row. Files must be saved before program execution.

How many observations in each table cell? 2
How many rows in your table? 5
How many columns in your table? 3
Which file contains column 1 ? fac1
Which file contains column 2 ? fac2
Which file contains column 3 ? fac3

```
           TABLE OF SUMS FOR EACH CELL

========================================================
Row   Col 1        Col 2        Col 3        Total:
--------------------------------------------------------
 1    162          158          142          462
 2    168          158          130          456
 3    184          162          170          516
 4    174          158          154          486
 5    152          144          124          420
--------------------------------------------------------
Tot:  840          780          720          2340


               A N O V A   T A B L E

=========================================================
Source of    Sum of                   Mean         F
Variation    Squares      d.f.         Square       Ratio
---------------------------------------------------------
Between Rows:   852        4            213          13.20

Between Cols:   720        2            360          22.31

Interaction:    224        8            28           1.74

Error:          242        15           16.13333
---------------------------------------------------------

Total:          2038       29           70.27586

Where now:    E=EXPLORE menu    R=Run ANOVA again    X=exit ? e
```

■ Exercises

4.1 A survey was taken in which twenty-eight randomly chosen college un-
dergraduate students were asked to state their grade point averages for the
previous semester. The results are shown in Table 4.6. Create four data

Table 4.6   Grade Point Averages of 28 Students

Freshman (5 students)	Sophomore (7 students)	Junior (10 students)	Senior (6 students)
1.91	3.88	3.01	3.26
2.44	2.02	2.86	2.45
3.17	2.96	3.45	2.81
2.25	3.52	3.67	3.02
2.71	2.59	3.33	3.02
	2.82	2.98	2.37
	3.11	3.26	
		3.44	
		4.00	
		3.08	

Table 4.7   Crash Damage in Dollars

Goliath	Varmint	Weasel
0	190	490
100	100	200
220	180	230
50	300	250
120	150	190

files named FRESH, SOPH, JUNIOR, and SENIOR, using the EX-PLORE file editor. Each data file should contain one column of numbers. Do a one-way analysis of variance to test the hypothesis that all four groups have the same mean grade point average. State your hypotheses clearly, and choose a level of significance. Use the computer program ANOVA for the calculations. Find the critical value of F from Appendix E, and illustrate your decision rule clearly. Discuss fully.

4.2 Three types of automobiles were deliberately crashed into a barrier at 5 mph, and the resulting damage (in dollars) was evaluated. Five test vehicles of each type were crashed. The results are shown in Table 4.7. Do a one-way analysis of variance to test the hypothesis that all the cars have the same mean dollar damage. State your hypotheses clearly, and choose a level of significance. Use the computer program ANOVA for the calculations, with three data files (GOLIATH, VARMINT, and WEASEL) typed in by using the EXPLORE file editor. Find the critical value of F using Appendix E, and illustrate your decision rule clearly. Discuss fully.

4.3 The waiting time (in minutes) for emergency room patients with non-life-threatening injuries was measured at four hospitals for all patients who arrived between 6:00 and 6:30 P.M. on a certain Wednesday. The results are shown in Table 4.8. Use the EXPLORE file editor to create four data files: HOSPA, HOSPB, HOSPC, and HOSPD. Use ANOVA to decide whether these sample results indicate that significant differences do exist

Table 4.8   Emergency Room Waiting Time in Minutes

Hospital A (5 patients)	Hospital B (4 patients)	Hospital C (7 patients)	Hospital D (6 patients)
10	8	5	0
19	25	11	20
5	17	24	9
26	36	16	5
11		18	10
		29	12
		15	

Table 4.9 Productivity of Assemblers in Plants

Plant A (9 workers)	Plant B (6 workers)	Plant C (11 workers)
3.6, 5.1, 2.8, 4.6, 4.7, 4.1, 3.4, 2.9, 4.5	2.7, 3.1, 5.0, 1.5, 2.2, 3.2	6.8, 2.5, 5.4, 6.7, 4.6, 3.9, 5.4, 4.9, 7.1, 8.4, 5.6

among the four hospitals, using the .05 level of significance. The director of one of the hospitals said, "These results prove nothing because the sample sizes are too small." Do you agree? Discuss fully.

4.4 Productivity measurements (average number of assemblies completed per minute) are taken for a random sample of workers at each of three plants. The tasks are comparable. The results are shown in Table 4.9. Use the EXPLORE file editor to create three data files called PLANTA, PLANTB, and PLANTC. Then use ANOVA to decide whether these sample results indicate the existence of a significant difference among plants at the .01 level. Discuss your results fully.

4.5 Three types of automobiles were deliberately crashed into a barrier at 5 mph, and the resulting damage (in dollars) was evaluated. Two test vehicles of each type were crashed. Crashes were made from three positions: head-on, slantwise, and rear-end. The results are shown in Table 4.10. Do a two-way analysis of variance to test the hypothesis that vehicle type makes a difference (column effect) or that crash angle makes a difference (row effect). Also, check for possible interaction effects. State all your hypotheses clearly, and choose a level of significance. Use the computer program ANOVA for the two-way ANOVA calculations, with three files (GOLIATH, VARMINT, and WEASEL), each file containing a column of numbers corresponding to this table. Find three critical F values using Appendix E, and illustrate your decision rules clearly. Discuss fully.

Table 4.10 Crash Damage in Dollars

Crash Type	Goliath	Varmint	Weasel
Head-on	0	200	380
	100	150	180
	50	130	240
Slant	200	350	500
	240	200	200
	140	150	340
Rear-end	0	0	150
	50	150	150
	0	50	50

Optional Exercises for Ambitious Learners

4.6 Choose two regions of the United States (for example, North, South). Find the unemployment rates for these two groups of states, and write them as two columns of numbers. Put each column in a file (such as NORTH and SOUTH). Alternatively, use program SPLIT to divide unemployment rates based on a binary variable of your own creation (or use NEAST, SEAST, or WEST). Run ANOVA, and do a one-way analysis of variance to test for significance difference of means. Then run TWOSAM on these same data. What differences do you find between the t test for two means and the F test for one-way ANOVA? At the same level of significance, do the tests agree? What would happen if you chose three regions instead of two (such as Northeast, Southeast, West)? Discuss fully.

4.7 Repeat Exercise 4.5, using the data matrix shown in Table 4.11 to create three new data files (if you use the same file names, the old files will be wiped out, so be sure you are finished with them first). In this data matrix there were six crashes (not three) of each type of vehicle and crash angle. It is an exact doubling of the data set in Exercise 4.5. What effect does doubling the sample size have on your test statistics? Discuss fully what this experiment tells you.

Table 4.11   Crash Damage in Dollars

Crash Type	Goliath	Varmint	Weasel
Head-on	0	200	380
	100	150	180
	50	130	240
	0	200	380
	100	150	180
	50	130	240
Slant	200	350	500
	240	200	200
	140	150	340
	200	350	500
	240	200	200
	140	150	340
Rear-end	0	0	150
	50	150	150
	0	50	50
	0	0	150
	50	150	150
	0	50	50

5

AREA/Finding Areas under Curves

Introduction

The program AREA calculates the right-tail area under any of four common continuous *probability density functions* (or p.d.f.'s) using the best available methods of approximation. For ease of interpretation, the areas are presented in the form of visual sketches. These sketches are not drawn to scale and are intended only to make it easier to visualize the area being estimated. The four models are

1. F,
2. Student's t,
3. Chi-square,
4. Normal.

These four continuous probability density functions are the most common ones used by statisticians for empirical testing of hypotheses. Using AREA, you can plug in your own sample test statistics (for example, from ANOVA, BIGRES, MGRES, or TWOSAM) to see just how significant your sample results are.

It is informative to compare the results with those found in Appendix B (normal), Appendix C (Student's t), Appendix D (chi-square), and Appendix E (F). Although the algorithms used here are generally good, some slight differences may be observed in extreme cases.

The converse situation, finding a critical value for a given area, is not so easy to calculate accurately and economically by using a microcomputer. However, there are programs for larger computers that do perform such calculations.

The right-tail normal area is evaluated by using an iterative integration of the normal probability density function. This method is highly accurate for the normal model because it does not involve any polynomial approximations.

Case Study 5.1: F, t, Chi-Square, and Normal Areas

The following printouts are self-explanatory. Each printout shows how **AREA** can be used to compute the right-tail area under a curve, using one of the four examples shown in Table 5.1. The examples in Table 5.1 are intended to be illustrative only and do not correspond to any particular empirical problem.

You can verify the accuracy of AREA for the four examples in Table 5.1 by using Appendixes B, C, D, and E. These appendixes will also be useful in setting up your own examples, as well as in other types of problems.

The program AREA uses the abbreviation "d.f." to mean *degrees of freedom*. The correct degrees of freedom will depend on your situation and cannot be generalized.

The computer methods used to evaluate areas for the first three models depend on degrees of freedom. For the F distribution, we use different approximations for odd and even degrees of freedom. For the Student's t in chi-square distributions, we use an exact method for small degrees of freedom but employ a normal approximation for large degrees of freedom. Further reading is suggested at the end of the chapter if you would like to know more.

Table 5.1 Four Right-Tail Areas under Probability Density Functions

Model and Example	Actual Right-Tail Area	Source of Actual Area
1. F distribution F = 5.74 Numerator d.f. = 30 Denominator d.f. = 4	.05	Appendix E
2. Student's t distribution t = 2.086 d.f. = 20	.025	Appendix C
3. Chi-square distribution χ^2 = 9.210 d.f. = 2	.01	Appendix D
4. Normal distribution z = 1.96	.025	Appendix B

```
                          AREA

    This program calculates approximate right-tail
    areas under several commonly used probability
    density functions.  Diagrams are provided only
    for heuristic illustration -- the shaded area
    is NOT proportional to the actual probability.

            PROBABILITY MODELS AVAILABLE:

                    1.   F
                    2.   Student's  t
                    3.   Chi-square
                    4.   Normal

    Input the number of the model you want (0 to quit)? 1
    Input F ? 5.74
    Input numerator d.f. ? 30
    Input denominator d.f. ? 4


    F PROBABILITY DENSITY FUNCTION

    |                  . . . . . .
    |                .              .
    |              .                  .
    |           .                      .
    |          .                       |.        Shaded Area =  0.050
    |         .                        |x.            d.f. =  30 , 4
    |                                  |xxx.
    |       .                          |xxxxxx.
    |.                                 |xxxxxxxxxxxxxx.
    +--------------------------------+-----------------
    0                                5.740



            PROBABILITY MODELS AVAILABLE:

                    1.   F
                    2.   Student's  t
                    3.   Chi-square
                    4.   Normal

    Input the number of the model you want (0 to quit)? 2
    Input Student's t? 2.086
    Input d.f.? 20

    STUDENT'S T PROBABILITY DENSITY FUNCTION

                         . . . . . .
                      .               .
                    .                   .
                   .                     .
                 .                       |.        Shaded Area =  0.0250
                .                        |x.            d.f. =  20
               .                         |xxx.
              .                          |xxxxxxx.
    ----------------------------------+---------+-----------------
                                      0        2.086
```

```
      PROBABILITY MODELS AVAILABLE:

          1.   F
          2.   Student's  t
          3.   Chi-square
          4.   Normal

Input the number of the model you want (0 to quit)? 3
Input Chi-Square? 9.210
Input d.f.? 2
```

CHI-SQUARE PROBABILITY DENSITY FUNCTION

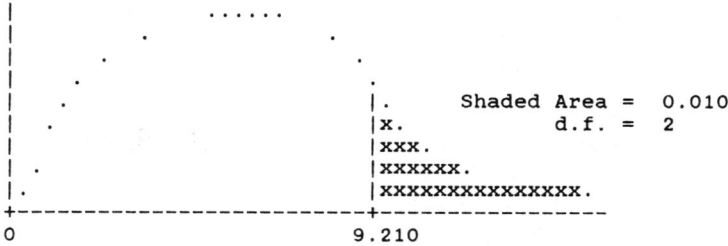

```
|                    . . . . . .
|                .                        .
|              .                        .
|            .                         .
|          .                         |.       Shaded Area =   0.010
|         .                          |x.            d.f. =   2
|                                    |xxx.
|        .                           |xxxxxx.
|      .                             |xxxxxxxxxxxxxxx.
|.                                   |xxxxxxxxxxxxxx.
+------------------------------------+-----------------
0                                   9.210
```

```
      PROBABILITY MODELS AVAILABLE:

          1.   F
          2.   Student's  t
          3.   Chi-square
          4.   Normal

Input the number of the model you want (0 to quit)? 4
Input z? 1.96
```

NORMAL PROBABILITY DENSITY FUNCTION

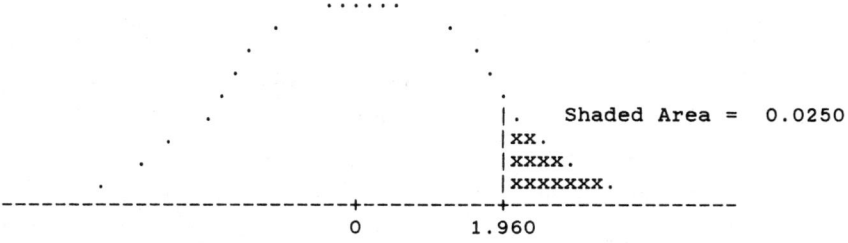

```
                      . . . . . .
                  .                 .
                .                     .
              .                         .
            .                             .
          .                             |.       Shaded Area =   0.0250
        .                               |xx.
      .                                 |xxxx.
    .                                   |xxxxxxx.
------------------------------------+----------+-----------------
                                    0        1.960
```

```
Input the number of the model you want (0 to quit)? 0

Where now:    E=EXPLORE menu    R=Run AREA again   X=exit ? e
```

■ Exercises

5.1 Use AREA to find the right-tail area under the F probability density function for the following cases:

a. F = 242 with d.f. = (10,1)

b. F = 19.4 with d.f. = (10,2)

c. F = 4.74 with d.f. = (10,5)

d. F = 2.98 with d.f. = (10,10)

e. F = 2.35 with d.f. = (10,20)

f. F = 1.91 with d.f. = (10,120)

Use Appendix E to verify that each of these areas *should* be exactly .05. What is your opinion of the accuracy of the algorithm used in AREA to approximate these F areas? What happens to the F value as denominator d.f. increases, holding the numerator d.f. constant?

5.2 Use AREA to find the right-tail area under the F probability density function for the following cases:

a. F = 4.96 with d.f. = (1,10)

b. F = 4.10 with d.f. = (2,10)

c. F = 3.33 with d.f. = (5,10)

d. F = 2.98 with d.f. = (10,10)

e. F = 2.77 with d.f. = (20,10)

f. F = 2.58 with d.f. = (120,10)

Use Appendix E to verify that each of these areas *should* be exactly .05. What is your opinion of the accuracy of the algorithm used in AREA to approximate these F areas? What happens to the F value as numerator d.f. increases, holding the denominator d.f. constant?

5.3 Use AREA to find the right-tail area under the Student's t probability density function for the following cases:

a. t = 6.314 with d.f. = 1

b. t = 2.920 with d.f. = 2

c. t = 2.015 with d.f. = 5

d. t = 1.812 with d.f. = 10

e. t = 1.725 with d.f. = 20

f. t = 1.660 with d.f. = 100

Use Appendix C to verify that each of these areas *should* be exactly .05. What is your opinion of the accuracy of the algorithm used in AREA to approximate these Student's t areas? What happens to the t values as d.f. increases? What limit, if any, are they approaching?

5.4 Use AREA to find the right-tail area under the chi-square probability density function for the following cases:

a. chi-square = 3.841 with d.f. = 1

b. chi-square = 5.991 with d.f. = 2

c. chi-square = 11.071 with d.f. = 5

d. chi-square = 18.307 with d.f. = 10

e. chi-square = 31.410 with d.f. = 20

Use Appendix D to verify that each of these areas *should* be exactly .05. What is your opinion of the accuracy of the algorithm used in AREA to approximate these chi-square areas? Are the chi-square values associated with the .05 area approaching a limit as d.f. increases?

5.5 Use AREA to find the right-tail area under the normal probability density function for the following cases:

a. z = 0.00

b. z = 1.00

c. z = 2.00

d. z = 3.00

e. z = 4.00

Use Appendix B to look up the correct areas. What is your opinion of the accuracy of the algorithm used in AREA to approximate these normal areas? Write a short paragraph stating what happens to the area under the normal curve as you move rightward from z = 0.00 to z = 4.00.

5.6 Use AREA to find the right-tail area under the normal probability density function for the following cases:

a. z = 1.282

b. z = 1.645

c. z = 1.960

d. z = 2.326

e. z = 2.576

What is special about each of these values, based on the areas you have found? Compare these values with those found in Appendix B. Then compare the z values with the t values found at the bottom line of Appendix C. What is the relationship, in your own words?

■ References

1. Kennedy, William J., Jr., and Gentle, James E., *Statistical Computing* (New York: Marcel Dekker, Inc., 1980), pp. 88–132.

2. Ling, Robert F., "A Study of the Accuracy of Some Approximations for t, χ^2, and F Tail Probabilities," *Journal of the American Statistical Association,* Vol. 73, 1978, pp. 274–283.

6

BIGRES/Bivariate Regression

Introduction

Bivariate regression provides a precise estimate of a proposed linear relationship between two quantitative variables. Regression is widely used because it offers a flexible way of answering all sorts of practical questions. For example, we might hypothesize that

$$\text{Quarterly sales} = f(\text{advertising expenditure})$$
$$\text{Telephone cost} = f(\text{number of calls})$$
$$\text{Lunch reimbursement} = f(\text{number in party}).$$

In each case a linear relationship might be a reasonable simplification of reality. Even though an *exact* linear relationship would not exist (because of a variety of other factors), an organization might be able to discover useful rules of thumb: How much *on the average* does it cost to take an extra client to lunch? To make an extra telephone call?

The program BIGRES estimates the equation of the best-fitting straight line for n pairs of observed data (X_i, Y_i) using the method of *ordinary least squares*. Data are read from two saved data files. In addition to the usual estimates, ANOVA table, computed t values, standard errors, and estimated Y values, BIGRES does a number of other useful things.

For example, BIGRES checks for awkwardly large or small data values. If any are present, BIGRES offers a warning about the need for rescaling. Most numerical results are rounded off and formatted, to simplify the presentation.

A label file (such as STATE or YEAR) may be used to help identify cases in the table of estimates and residuals.

Visual displays and tests of the residuals are featured prominently to enhance the user's awareness of possible problems with non-normality, autocorrelation, or heteroskedasticity.

PLOT may be used along with BIGRES to see the relationship, if any, between X and Y. More complex models (with more than two variables) will require MGRES (multiple regression).

Case Study 6.1: Estimating a Tax Function

BIGRES can be used to estimate the relationship between personal taxes and personal income. Table 6.1 shows aggregate data for the period 1960–1985, as stored in the time-series database files PI (personal income), PT (personal taxes), and YEAR.

A hand-drawn *scatter diagram* (Figure 6.1) suggests a strong linear relationship. This relationship is partly definitional (reflecting merely the upward trends in the two variables).

After looking at the scatter diagram (which could be obtained automatically by using PLOT, if desired) it seems reasonable to try to fit a straight line to the observed data. The estimated line will have the form

$$Y = a + bX$$

where

Y = aggregate personal taxes (billions of current dollars)

X = aggregate personal income (billions of current dollars).

When we run BIGRES, the computer asks for the file names (PT and PI) and a label file (YEAR). Next, a table of means and sample standard deviations lets us verify that our files have the correct length and magnitude. BIGRES then prints a table of sums, sums of squares, and sums of cross-products. This table would permit us to verify the calculations of the estimated line of regres-

**Table 6.1   U.S. Personal Income and
Personal Taxes, 1960–1985**

Year	Income	Taxes	Year	Income	Taxes
1960	409.4	50.5	1973	1101.7	152.0
1961	426.0	52.2	1974	1210.1	171.8
1962	453.2	57.0	1975	1313.4	170.6
1963	476.3	60.5	1976	1451.4	198.7
1964	510.2	58.8	1977	1607.5	228.1
1965	552.0	65.2	1978	1812.4	261.1
1966	600.8	74.9	1979	2034.0	304.7
1967	644.5	82.4	1980	2258.5	340.5
1968	707.2	97.7	1981	2520.9	393.3
1969	772.9	116.3	1982	2670.8	409.3
1970	831.8	116.2	1983	2836.4	411.1
1971	894.0	117.3	1984	3111.9	441.8
1972	981.6	142.0	1985	3294.2	493.1

Source: *Economic Report of the President* (Washington, D.C.: U.S. Government Printing Office, 1986), p. 282. Figures are in billions of current dollars.

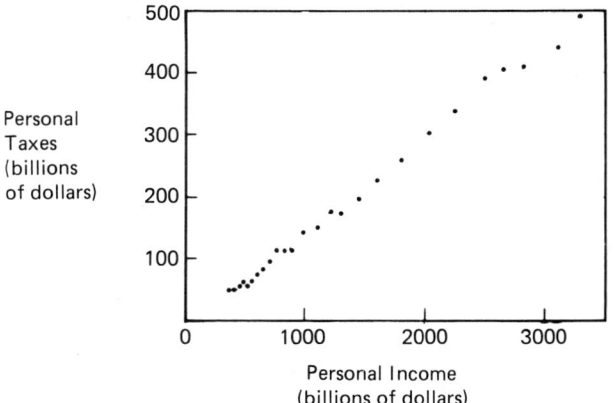

Figure 6.1 **Hand-Drawn Scatter Diagram of Taxes versus Income**

sion, using the ordinary least squares formulas (see your statistics text for these formulas).

Rounding things off a bit, BIGRES's estimated *line of regression* may be written as

$$\hat{Y} = -14.54 + .1535X.$$

This is an estimate of the underlying bivariate regression model:

$$Y = \alpha + \beta X + \text{random error}.$$

The *estimated slope* is

$$b = .1535 = \frac{dY}{dX} = \text{rate of change in taxes with respect to income}.$$

Thus a one-dollar increase in income leads to about fifteen cents of extra taxes, on the average. In other words, the *effective marginal tax rate* is about fifteen cents on the dollar. Of course, it is unrealistic to think that a simple bivariate model will suffice for economic analysis, since there are many other variables and models to be considered before claiming that we have measured *the* marginal tax rate. But this is a reasonable first exploratory step toward such an estimate.

The *estimated intercept* is approximately −14. This means that if personal income were zero, taxes would be negative—the government would be paying money out, instead of taking it in! This scenario makes little sense and is well beyond the range of observed data, so the intercept is not of much empirical interest.

BIGRES

This is a bivariate regression program to estimate
the linear model Y = A + B X using ordinary least squares.
Program expects two saved data files (one file per variable).
Each file may contain up to 500 cases.

Which file is Y (dependent variable) ? pt
Which file is X (independent variable)? pi
Want to use a label file (y/n)? y
Name of label file? year

File	Cases	Mean	St. Dev.
PI	26	1364.735	915.1019
PT	26	194.8885	140.7139

TABLE OF SUMS
```
=============================
```
File for X	PI
File for Y	PT
Sum X	35483.1
Sum Y	5067.1
Sum X*X	6.93603E+07
Sum Y*Y	1482529
Sum X*Y	1.012791E+07
Mean of X	1364.735
Mean of Y	194.8885

Sum $(X-\bar{X})*(X-\bar{X})$ 2.093529E+07

Sum $(Y-\bar{Y})*(Y-\bar{Y})$ 495009.8

Sum $(X-\bar{X})*(Y-\bar{Y})$ 3212661

Estimated Line of Regression:
  Y = -14.53928 + .153457 X

A N O V A   T A B L E

Source	Sum of Squares	D.F.	Mean Square	F
Regression	493004.5	1	493004.5	5900.205
Error	2005.372	24	83.55718	
Total	495009.8	25		

BIGRES's *analysis of variance* (ANOVA) table shows that the sum of squares due to regression (SSR) is quite large in relation to the error sum of squares (SSE). This leads to a very large test statistic when we take the ratio of the mean squares:

$$F = \frac{SSR/(1)}{SSE/(n-2)} = \frac{493004.5/1}{2005.372/24} = \frac{493004.5}{83.55718} = 5900.205.$$

We hardly require a formal hypothesis test to know that this huge F statistic is highly significant. Consulting Appendix E, we find that the critical value at the .01 level of significance is $F = 8.10$, using d.f. $= 1,20$. (Since d.f. $= 1,24$ is not in the table, we use the next lower degrees of freedom to be conservative about committing Type I error.) A large F statistic means that there is a strong relationship between X and Y. In fact, our F statistic is so large that we become suspicious. We would not expect to see such large F statistics in most research applications. We make a mental note to check the underlying assumptions carefully later on and to look for sources of spurious correlation.

```
SUMMARY OF FIT OF REGRESSION OF PT ON PI

Estimated standard error of slope =  .001998
Estimated standard error of intercept =  3.263032

Computed t value for slope     =  76.813
Computed t value for intercept =  -4.456

Correlation coefficient = 0.9980
R-Squared = 0.9959
Per cent of variation explained =  99.59
Estimated standard error of estimate = 9.140962
```

Before doing a test for significance on the slope, we might want to construct confidence intervals for both parameters of the regression line, using their estimated standard errors. BIGRES reveals that

$$\hat{\sigma}_a = 3.263 \qquad \hat{\sigma}_b = .001998.$$

Using the fact that a two-sigma range corresponds to about 95 percent of the area under the student's t distribution, these standard deviations yield "quick" 95-percent confidence intervals approximately as follows:

$$b \pm 2\hat{\sigma}_b \rightarrow \quad .1535 \pm 2\,(.001998) \quad \rightarrow \quad .150 < \beta < .157$$
$$a \pm 2\hat{\sigma}_a \rightarrow \quad -14.54 \pm 2\,(3.263) \quad \rightarrow \quad > -21.1 < \alpha < -8.0.$$

Although these results would have to be refined, they do make it clear that the slope estimate is very precise (narrow confidence interval), while there is more uncertainty about the true intercept (wide confidence interval).

Based on the *a priori* belief that the slope should be positive, we can test these hypotheses:

$$H_0: \quad \beta = 0$$
$$H_1: \quad \beta > 0.$$

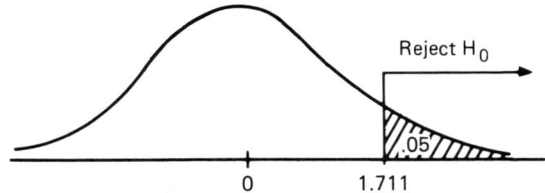

Figure 6.2 Decision Rule for the Slope Using Student's t

In a right-tail test, using d.f. = n − 2 = 26 − 2 = 24, Appendix C yields a critical value of student's t (1.711) for the .05 level of significance. The decision rule is shown in Figure 6.2. From BIGRES the test statistic for the slope is t = 76.813, which easily leads to the rejection of H_0. The BIGRES calculation (except for slight rounding error) is as follows:

$$t = \frac{b}{\hat{\sigma}_b} = \frac{.153457}{.001998} = 76.805.$$

Therefore the slope is significantly greater than zero and cannot be attributed to chance.

Caution is needed, though. First, this large t statistic may overstate the relationship between PT and PI because of *autocorrelation,* a common phenomenon in time-series data. Second, the relationship between PT and PI is partly attributable to a strong upward *time trend* in both variables.

BIGRES's calculated *correlation coefficient* (r = .9980) is extremely close to its theoretical maximum (r = 1.0000), signifying a near-perfect positive correlation. A t test of the correlation coefficient shows

$$t = r\sqrt{\frac{n - 2}{1 - r^2}} = .9980\sqrt{\frac{26 - 2}{1 - .9980^2}} = 77.343.$$

Except for roundoff (in printing r = .9981), this calculation gives the same result as the test for the slope. The apparent inaccuracy arises because the denominator in the t calculation is very close to zero. BIGRES maintains more accuracy than is shown in this calculation, so you can rely on the computer printout.

The term r^2, called the *coefficient of determination,* is probably the most commonly used measure of fit in regressions. For our data, BIGRES's r^2 = .9959 indicates that 99.59 percent of the variation in taxes is explained by income. This apparently strong result should be viewed with caution: High coefficients of determination, although rare in social sciences such as psychology or business disciplines such as organizational behavior and marketing, are common in economic data. Here, the r^2 mainly reflects the upward trend of both variables over time. If we were to remove trend from each variable and rerun BIGRES, we would expect the r^2 to fall considerably. Such a procedure would be needed to eliminate the *spurious correlation* due to trend alone. Nonetheless, all else being equal, a good fit is preferable to a poor fit. BIGRES's formula for r^2 is

$$r^2 = 1 - SSE/SST$$

where

SSE = sum of squared errors about the estimated Y

SST = sum of squared deviations about the mean of Y.

If the sum of squared errors is near 0, it is clear that r^2 will be close to its maximum value of 1 (as in our BIGRES run), signifying an excellent fit. Conversely, if the regression's fit is poor, SSE will be almost as large as SST (SSE cannot exceed SST), and hence r^2 will be close to its minimum value of 0. In practice, any r^2 value between 0 and 1 can occur.

The *estimated standard error of the regression* (or simply *standard error*) also offers insight into the fit of the model. The standard error is measured in the same units as the dependent variable (billions of dollars). The formula used by BIGRES is

$$SE = \sqrt{\frac{\Sigma(Y_i - \hat{Y}_i)^2}{n - 2}} = 9.141.$$

The two-sigma range may be used to assess the approximate width of a 95-percent confidence interval for predicting *individual* Y values:

$$\hat{Y} \pm 2SE$$
$$\hat{Y} \pm 2(9.141)$$
$$\hat{Y} \pm 18.282.$$

So we can say that our prediction interval's width is about plus-or-minus 18 billion dollars. Clearly there is room for improvement, despite the very high coefficient of determination. Adjustments could be made to widen the confidence interval when X is far above or below its mean, and t-values from Appendix C could be used instead of the crude two-sigma range. But this quick calculation gives an idea of our regression's accuracy.

If we wish to predict only the *mean* of Y, a much narrower approximate 95-percent confidence interval may be constructed by dividing SE by the square root of the sample size:

$$\hat{Y} \pm 2SE/\sqrt{n}$$
$$\hat{Y} \pm 2(9.141)/\sqrt{26}$$
$$\hat{Y} \pm 3.585.$$

These approximate 95-percent confidence intervals actually should use t values from Appendix C for $n - 2$ degrees of freedom. But the assumption that $t = 2.000$ is a reasonable "quick" approximation unless the sample size is tiny. For other levels of confidence, Appendix C can be used.

The *residuals* (differences between actual and estimated Y values) give very important clues about the model's fit on a year-by-year basis. We should look closely at cases in which the model fits poorly to get ideas for adding new predictors if we later decide to build a multivariate regression model. The table of residuals contains five years that stand out because of their unusually large residuals: Large positive residuals (unusually high taxes) are observed in 1969, 1981, and 1982, and large negative residuals (unusually low taxes) are observed in 1975 and 1984.

The explanation for instances of poor fit can be found by studying tax legislation approved by Congress and the timing of business recessions. Perhaps these noneconomic factors could be incorporated into a multiple regression model, using binary variables or some other method.

What is the definition of a *large* residual? In this case a simple inspection of the BIGRES printout suggests singling out those residuals that exceed 10 billion dollars, as we have done. A somewhat casual taxonomy could be devised, such as

Outlier → residual exceeds 3.0 times the standard error

Large → residual exceeds 2.0 times the standard error.

Noteworthy → residual exceeds 1.0 times the standard error.

ACTUAL AND ESTIMATED PT VALUES

YEAR	X	Y Actual	Y Estimate	Residual
1960	409.4	50.5	48.28592	2.21408
1961	426	52.2	50.83330	1.36670
1962	453.2	57	55.00732	1.99268
1963	476.3	60.5	58.55217	1.94783
1964	510.2	58.8	63.75436	-4.95436
1965	552	65.2	70.16885	-4.96885
1966	600.8	74.9	77.65754	-2.75754
1967	644.5	82.4	84.36359	-1.96359
1968	707.2	97.7	93.98534	3.71466
1969	772.9	116.3	104.06745	12.23255
1970	831.8	116.2	113.10604	3.09396
1971	894	117.3	122.65106	-5.35106
1972	981.6	142	136.09386	5.90614
1973	1101.7	152	154.52402	-2.52402
1974	1210.1	171.8	171.15874	0.64127
1975	1313.4	170.6	187.01082	-16.41081
1976	1451.4	198.7	208.18785	-9.48785
1977	1607.5	228.1	232.14244	-4.04243
1978	1812.4	261.1	263.58572	-2.48572
1979	2034	304.7	297.59174	7.10828
1980	2258.5	340.5	332.04279	8.45721
1981	2520.9	393.3	372.30981	20.99017
1982	2670.8	409.3	395.31302	13.98697
1983	2836.4	411.1	420.72543	-9.62543
1984	3111.9	441.8	463.00275	-21.20276
1985	3294.2	493.1	490.97794	2.12207

While such a rule would correspond to what practicing statisticians actually do, it is only a rule of thumb and not a precise law of statistics. It is suggested only as a convenient guide to scanning a printout so that your attention is focused on the unusual cases. It may seem paradoxical that we ignore the cases in which the model fits well, but scientific progress is often made by studying exceptions.

```
    ....   TEST 1: NORMALITY OF RESIDUALS   ....

   f(o)      f(e)     z Scale and Residual Histogram
  ------    ------      +------------------------------
    0        0.0       |
                     -3+
    1        0.6       | *
                     -2+
    3        3.5       | ***
                     -1+
    8        8.9       | *******
                      0+
   11        8.9       | **********
                      1+
    2        3.5       | **
                      2+
    1        0.6       | *
                      3+
    0        0.0       |

      Chi-Square =  0.906 with d.f. =  1

 First and last 3 classes were collapsed to enlarge f(e).
```

BIGRES's histogram of residuals does not look bell-shaped—there seem to be too many asterisks clustered near the middle. However, there are no true outliers (beyond three standard deviations). Formally, we can test the hypotheses:

H_0: Errors are normally distributed.

H_1: Errors are not normally distributed.

The decision rule is shown in Figure 6.3, using Appendix D with d.f. = 1 and the .05 level of significance to obtain a critical value of chi-square.

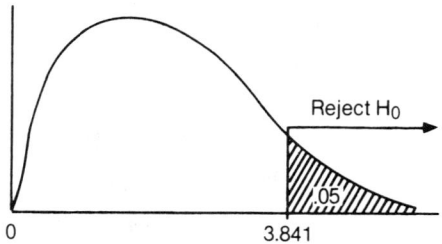

Figure 6.3 Decision Rule for Normality of Residuals

We cannot reject H_0, since the test statistic (.906) does not exceed the critical value (3.841). However, we are still left with the feeling that the residuals are not quite normal based on an "eyeball" inspection of the histogram. Non-normal residuals would not pose any serious hazard to our regression unless the sample size were quite small. At worst they might make the confidence intervals unreliable.

```
.... TEST 2: AUTOCORRELATION OF RESIDUALS ....

Correlation of e(i) with e(i-1) = 0.393
Computed t for autocorrelation is 2.051 with d.f. = 23
Durbin-Watson test statistic = 1.21
```

We expect *autocorrelation* (pattern in the residuals) in a time-series model, and so we are not surprised by a positive estimated autocorrelation coefficient (.393) that appears too far away from 0 to be due to chance. This conclusion may be checked by a formal two-tailed test of the hypotheses:

H_0: No autocorrelation exists.

H_1: Autocorrelation is present.

Using Appendix C with d.f. = 23 and the .05 level of significance, we obtain the decision rule shown in Figure 6.4. Since the test statistic (2.051) does not quite exceed the critical value (2.069) of student's t, we cannot reject H_0. However, the decision is very close. A different level of significance or a right-tail test could have led to the rejection of H_0.

Statisticians often rely on the Durbin-Watson statistic as a measure of autocorrelation. Ideally, the Durbin-Watson test statistic should be close to 2. Since BIGRES's calculated Durbin-Watson test statistic (1.21) is noticeably less than 2, we suspect positive autocorrelation. (The opposite, a value greater than 2, would be unusual in economic data.) Complete tables for the Durbin-Watson test may be found in more advanced texts.

Inspecting BIGRES's column of residuals, we can see a pattern of several +'s followed by several −'s, which would indicate positive autocorrelation. This reinforces our feeling that autocorrelation may exist.

Overall, it seems prudent to regard autocorrelation as a potential hazard to our model. Autocorrelation will not cause our estimates to be biased, but

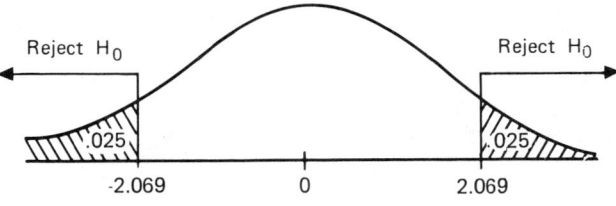

Figure 6.4   Test for Autocorrelation Using Student's t

it could cause some loss of efficiency in the estimates and may cause the t-value for the slope to be overstated, possibly misleading us about the strength of the association between PT and PI. As long as it is mild, we do not worry about it. But in our example it could be a danger.

```
.... TEST 3: HETEROSKEDASTICITY OF RESIDUALS ....

==========================================
Group  Cases        Variance of Residuals
------------------------------------------
1        8                    9.40371
2        9                   66.95989
3        9                  147.50043
------------------------------------------

Bartlett's Chi-Square = 5.069 with d.f. =  2

Where now:   E=EXPLORE menu     R=Run BIGRES again     X=exit ? e
```

The last check of the model's assumptions using BIGRES is for possible *heteroskedasticity* (nonconstant error variance). From BIGRES's table of grouped residual variances we see that Group 3 has a larger residual variance (147.50) than does Group 1 (9.40). Ideally, the residual variances would be the same, regardless of how the groups are formed. This apparent inequality of residual variances could signal another violation of the classic regression assumptions. We can use these grouped variances to perform an approximate F test called *Hartley's test* of the hypotheses:

H_0:   Errors are homoskedastic (constant variance).

H_1:   Errors are heteroskedastic (changing variance).

An approximate F test statistic is

$$F = \frac{\text{largest residual variance}}{\text{smallest residual variance}} = \frac{147.50}{9.40} = 15.69.$$

Degrees of freedom for this test will depend on the number of groups (c) and the sample size (n) as follows:

Numerator d.f. = c = 3.

Denominator d.f. = n/c − 1 = 26/3 − 1 = 8 − 1 = 7.

At the .05 level of significance, Appendix E yields a critical value (4.07) that is far less than the observed F ratio (15.69), so we reject H_0 and conclude that significant heteroskedasticity over time is present.

Another way to test these hypotheses about homoskedasticity is with Bartlett's test. Using the .05 level of significance, with d.f. = c − 1 = 2, we can set up the chi-square decision rule illustrated in Figure 6.5. Since Bartlett's test statistic (5.069) is less than the critical value of chi-square (5.991), we cannot reject H_0. But we are very close to the rejection region.

Bartlett's test is known to be somewhat insensitive in cases where the residuals are non-normal. A simple eyeball inspection of BIGRES's table of

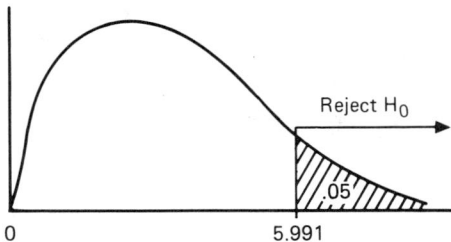

Figure 6.5   Bartlett's Test for Heteroskedasticity

residual variances or of the scatter plot of Y against X is probably the most reliable way to detect heteroskedasticity. Heteroskedasticity is sometimes unavoidable in time-series data with strong upward trend and causes less concern than in cross-sectional data. Even if heteroskedasticity is detected, the estimates will still be unbiased, though there may be some loss of efficiency.

If we are using cross-sectional data we should sort Y on X before running BIGRES, to ensure valid tests for heteroskedasticity. Unless this is done, the results of the heteroskedasticity tests will merely reflect the order of data entry. Presorting Y on X can be accomplished using SORTER. With time series data, we would usually retain the original order. Ideally, a regression should have residuals that are normally distributed (bell-shaped), nonautocorrelated (no pattern over time), and homoskedastic (constant variance over all X values). Violations of the latter two conditions are considered more serious. But the estimates remain unbiased under a rather wide range of conditions. You should not let all these tests worry you into thinking that your model can be totally invalidated by a minor problem. Most real-world regressions do exhibit slight violations of these assumptions and still remain useful.

We might be able to reduce autocorrelation, and perhaps alleviate other problems, by rerunning this time-series model with new variables defined as period-to-period changes (also known as *first differences*):

$$\Delta X = \text{the change in personal income}$$
$$\Delta Y = \text{the change in personal taxes.}$$

Although the fitted relationship will be less significant, potential residual problems may be lessened. Other approaches, such as variable transformations, might also be tried.

■ Technical Notes for Advanced Students

The most common transformation when *first-order autocorrelation* is a problem is called *generalized differences*. This method involves replacing X and Y with new variables X' and Y', defined as follows:

$$\left. \begin{array}{l} X'_t = X_t - \hat{r}X_{t-1} \\[2mm] Y'_t = Y_t - \hat{r}Y_{t-1} \end{array} \right\} \quad \text{for } t = 2, 3, \ldots, n,$$

where $\hat{r}$ is the *estimated coefficient of autocorrelation* (obtained from BIGRES's test 2). A special transformation is used for the first observation:

$$X'_1 = X_1 \sqrt{1 - \hat{r}^2}$$
$$Y'_1 = Y_1 \sqrt{1 - \hat{r}^2}.$$

We ignore the possibility of negative autocorrelation ($\hat{r} < 0$), since it is of less logical interest and does not appear in time-series data very often. Assuming that $\hat{r}$ lies between 0 and 1, we may say that in general the new variables are formed by subtracting from each observation a fraction of its previous value. If $\hat{r} = 0$ (no autocorrelation), there is no carryover from one period to the next, so the new variables are the same as the old. At the other extreme, if $\hat{r} = 1$, this transformation amounts to using first differences (that is, each variable is replaced by its *change* over the previous period).

This transformation may be made by using the program TRANS (see Chapter 24). The TRANS printout is not shown, but the transformation may be explained by partially listing the old files (PT and PI) and new files (NEWPT and NEWPI) generated by the program TRANS:

t	PT	NEWPT	PI	NEWPI
1	50.5	46.43670	409.4	376.4591
2	52.2	32.35350	426.0	265.1058
3	57.0	36.48540	453.2	285.7820
.	.	.	.	.
.	.	.	.	.
.	.	.	.	.
25	441.8	280.23767	3111.9	1997.1947
26	493.1	319.47260	3294.2	2071.2231

For example, if we use the old variable PT and the estimated autocorrelation coefficient $\hat{r} = .393$ from BIGRES (see Test 2 for autocorrelation), the calculations would look like this:

$$Y'_1 = Y_1 \sqrt{1 - \hat{r}^2} = 50.5 \sqrt{1 - .393^2} = 46.44$$
$$Y'_2 = Y_2 - \hat{r}Y_1 = 52.2 - .393 (50.5) = 32.35$$
$$\cdot$$
$$\cdot$$
$$\cdot$$
$$Y'_{26} = Y_{26} - \hat{r}Y_{25} = 493.1 - .393 (441.8) = 319.47.$$

The calculations by TRANS are automatic, so all you really need is the estimated autocorrelation coefficient from the residual tests of BIGRES. (Caution: be sure not to confuse it with the overall correlation coefficient for the model.)

We call our new variables NEWPT and NEWPI. When we rerun BIGRES, we obtain

$$Y' = -9.44281 + .15399X'$$
$$(-3.114) \qquad (53.051)$$

The estimated slope is about the same as before (it was previously estimated to be .1535), and the computed t-statistics (in parentheses) show that the slope is still significantly different from zero, although its estimated standard error is larger and its t-value is lower than before. After the transformation the overall r^2 drops to .9915 but is still quite high. The new residuals reveal an estimated autocorrelation coefficient of only .135, with a Durbin-Watson statistic of 1.68, which comes much closer to the ideal.

Heteroskedasticity persists in the new model. Thus while variable transformations help, they do not solve all our problems. Probably, a more complex model should be specified, rather than further refining this bivariate relationship.

■ Exercises

6.1 From the DATABASE-1 diskette, choose two state data files or two time-series files (but not one of each). The emphasis should be on choosing one variable that you wish to "explain" and one independent variable (predictor) that might reasonably be expected to exhibit a relationship with your dependent variable.

6.2 Run LAYOUT to obtain a listing of your two data files for the regression. Also list the correct file to label each variable (STATE if you are using the state data files and YEAR if you are using the time-series files). Discuss fully how you imagine the data were collected, including possible limitations in accuracy (such as measurement error, unreliable sources, and so on). Discuss the units of measurement and make sure the data are well-conditioned. If they are not, transform them as necessary using TRANS. If you are using cross-sectional data (e.g., state database) you may wish to use SORTER to presort Y on X to guarantee valid tests for heteroskedasticity. If so, save the sorted observations under new file names, to avoid confusion. Presorting will only affect the tests for autocorrelation and heteroskedasticity.

6.3 Before analyzing the data, discuss your *a priori* reasoning about cause-and-effect, probable signs of the correlation coefficient and slope, and how close a fit you would expect. Write down these thoughts, including your predictions about signs, in readable essay format.

6.4 Use the computer program PLOT to obtain a scatter diagram. Sketch in a regression line by eye. Try to guess the correlation coefficient visually—without doing any calculations. Just remember that a correlation coefficient must lie somewhere between a perfect inverse correlation, or downward-sloping straight line (correlation coefficient $= -1.000$), and a perfect positive correlation, or upward-sloping straight line (correlation coefficient $= +1.000$). A complete lack of relationship would correspond to a scatter plot whose dots form no particular pattern. Try drawing your own scatter diagram by hand, trying to improve on the computer's.

6.5 Run the computer program BIGRES to obtain all available regression statistics. Check the table of X and Y to verify that your data files are correct. (See Appendix H for the correct data values.) Use a label file (such as STATE or YEAR) to identify your cases. If you presorted Y on X, be sure your label file is also presorted on X.

6.6 Discuss how close your visually estimated correlation coefficient was to the actual. Without doing formal tests, discuss the overall fit using various criteria. Be specific about what each statistical measure of fit tells you and what overall strength you believe exists in the relationship between X and Y. Do you believe that the straight-line model is a reasonable approximation?

6.7 Write the equation of the estimated line of regression. Discuss the meaning of the slope: What does it say about the effect of X upon Y? What meaning, if any, does the intercept have?

6.8 Set up a pair of hypotheses about the slope and test them, using Appendix C to obtain the critical values of student's t. Do the same for the intercept. Choose your own level of significance and illustrate your decision rules. State why you chose a one- or two-tail test. Are your decisions sensitive to the choice of level of significance? Discuss all your conclusions fully.

6.9 Set up a pair of hypotheses about the true correlation coefficient and test them, using the same level of significance as in Exercise 6.8. Use Appendix C to obtain the critical value of student's t. Is the decision sensitive to the choice of significance level? Illustrate and discuss.

6.10 Discuss your *a priori* predictions. If they were not borne out, discuss reasons why the expected relationship might not be detectable in a bivariate regression.

6.11 Construct a 95-percent confidence interval for a particular Y value, choosing any X as the point of departure. Compare the "quick rule" (which assumes that $t = 2$ and multiplies by the estimated standard error, which is assumed to be constant) and the "better" method (which uses t for d.f. $= n - 2$ from Appendix C and multiplies by the estimated standard error, which is assumed to be constant). Discuss the practical differences, if any. Is your actual Y within these confidence intervals? What difference would it have made if you needed a 90-percent confidence interval instead of 95 percent?

6.12 Study BIGRES's table of residuals. On your printout, circle in red any residuals exceeding three standard errors. Circle in blue any residuals exceeding two standard errors. Circle in pencil any residuals exceeding one standard error. Then jot down beside each of these large residuals the name of the corresponding state (or year). In two columns, list the state names (or years) corresponding to these large residuals: positive residuals in column one, negative residuals in column two. What generalization can you make about the states (or years) that were underestimated (column one) and overestimated (column two) by your model? Suggest possible reasons why the model might not predict well in those cases. Could you suggest other predictors besides your X that might be tried?

6.13 What percent of your residuals are actually within the limits of your quick (two-sigma) 95-percent confidence interval (refer to Exercise 6.11)? This question may easily be answered by just counting the red and blue circled residuals (see Exercise 6.12) and subtracting from n. If the answer differs substantially from 95 percent, discuss.

6.14 Perform appropriate tests for normality of the residuals. Discuss the appearance of the histogram, telling whether there are outliers and then doing formal tests. Are your conclusions sensitive to the level of significance? Illustrate your decision rules. Discuss the implications of your conclusions about normality.

6.15 Perform appropriate tests to detect autocorrelation. Illustrate your decision rule(s). Can you actually see any pattern in the column of residuals? What is the implication, if any, for your model?

6.16 Perform appropriate tests to detect heteroskedasticity of the residuals. Discuss the differences, if any, between the visual check of residual variances in the groups, the Hartley's F test, and Bartlett's test. Illustrate your decision rules. What is the implication for your model? Was presorting Y on X necessary? Why, or why not?

6.17 Statistics is supposed to help understand the behavior of people, things, and numerical measures of human activities. How has your understanding changed as a result of this assignment? Is the computer a friend or a foe in your work?

■ References

Wonnacott, R. J. and Wonnacott, T. H., *Econometrics,* 2nd edition (New York: John Wiley and Sons, 1979), pp. 215–219.

7

BINOM/Binomial Probability Model

Introduction

The program BINOM calculates exact and cumulative *binomial probabilities* using the binomial p.d.f. (probability density function):

$$P(x) = \frac{n!}{x!(n - x)!} \, p^x \, (1 - p)^{n-x}$$

where

$$x = \text{number of successes}$$
$$n = \text{number of independent trials}$$
$$p = \text{probability of success in each trial.}$$

For example, the probability of 9 heads out of 10 tries is

$$P(9) = \frac{10!}{9!1!} \, (.5)^9 \, (1 - .5)^1 = .0098.$$

Since binomial events are mutually exclusive, we can apply the additive law of probabilities. For example,

$$P(9 \text{ or } 10) = P(9) + P(10)$$
$$= .0098 + .0010$$
$$= .0108.$$

BINOM suppresses printing of very small probabilities. BINOM uses logarithms to preserve accuracy over a wide range of n and p values. Factorials are so unwieldy that it is no fun to attempt any but the simplest binomial probabilities on a calculator (if you are not convinced, try taking 79! on your calculator).

As n increases, any binomial p.d.f. approaches a normal, bell-shaped curve; and when n is large, it becomes appropriate to use a normal approximation. For such an approximation we would rely on the binomial experiment's mean, variance, and standard deviation, which are as follows:

$$E(X) = np$$
$$V(X) = np(1 - p)$$
$$S.D. = \sqrt{np(1 - p)}.$$

If n is large and p is small, it may be appropriate to use a Poisson approximation with mean np. If a normal or Poisson approximation would be desirable, BINOM will say so and will give suggestions for setting up the approximation.

BINOM also computes the skewness and kurtosis coefficients for whatever n and p you are using. The formulas are

$$\text{Skewness} = (1 - 2p)/\sqrt{np(1 - p)};$$
$$\text{Kurtosis} = 3 - 6/n + 1/[np(1 - p)].$$

When $p = .5$, the skewness coefficient will be zero; otherwise, the skewness will be either positive (if $p < .5$) or negative (if $p > .5$). But as n increases, the skewness approaches zero (assuming that p is held constant). Sketch the appearance of the binomial for a large n (for any p), and you will see the intuitive validity of this fact.

Regardless of p, the kurtosis will generally differ from 3 (recall that a normal p.d.f. has a kurtosis coefficient of 3). However, as n increases, the last two terms in the kurtosis formula approach zero (holding p constant), so the kurtosis coefficient approaches 3. This supports the idea that a binomial p.d.f. approaches the shape of a normal curve as n increases.

◼ Case Study 7.1: Binomial with Large n and Small p

The capabilities of BINOM may be best illustrated by choosing a calculation that would be quite difficult without a computer, namely, a table of binomial probabilities using a large number of trials. Let us choose

$$n = 1000$$
$$p = .001.$$

Although this calculation can readily be approximated by using the Poisson model with a mean equal to

$$np = (1000)(.001) = 1,$$

it would be a bit difficult with a hand calculator because you would have to calculate

$$P(x) = \frac{1000!}{x!(1000 - x)!} .001^x (1 - .001)^{1000 - x}.$$

However, this calculation poses no difficulty for BINOM. As the sample print-out shows, the computer even recommends the Poisson alternative, which you could do with the program POIS (or with a table in a statistics book).

This is not to imply that BINOM can do *any* calculation. But if the task would be too time-consuming or would suffer a grievous loss of accuracy, BINOM will refuse to perform. So go ahead and try any experiments you wish, and see what happens.

In this example,

$$\text{Skewness} = (1 - 2p)/\sqrt{np(1 - p)}$$
$$= [1 - 2(.001)]/\sqrt{1000(.001)(.999)}$$
$$= .998$$

and

$$\text{Kurtosis} = 3 - 6/n + 1/[np(1 - p)]$$
$$= 3 - 6/1000 + 1/[1000(.001)(.999)]$$
$$= 3.995.$$

Thus the p.d.f. is skewed to the right and leptokurtic (more peaked than normal). A sketch of the p.d.f. (from the BINOM printout) will verify this shape. Even though the sample is large, the p.d.f. still is not very "bell-shaped" because p is so small. Had we chosen p nearer .5, the resemblance to a normal curve would have been more obvious.

```
                            BINOM

    This program calculates and prints Binomial probabilities:

            1. Exact Binomial probabilities
            2. Right-tail cumulative Binomial probabilities
            3. Left-tail cumulative Binomial probabilities

    You will be asked to provide two Binomial model parameters:

                n = number of trials
                p = probability of success on each trial

    Printing is suppressed if probabilities become very small
    in order to shorten the table.

    How many trials (n)? 1000
    What is the probability of a success (p)? .001
    For this model, skewness =  0.998 and kurtosis = 3.995


    AN ALTERNATIVE TO CONSIDER

    Since your p is small (n = 1000  and p = .001 ) you might also
    want to try the Poisson approximation with:

            Mean  =  1  =  n*p
```

```
            BINOMIAL PROBABILITY TABLE

    n =  1000  = number of tries
    p =  .001  = probability of success on each try
    r = number of successes in 1000 tries

              Exact        Cumulative      Cumulative
      r       P(X=r)       P(X>=r)         P(X<r)
    ----      ------       -------         ------
      0       0.3677       1.0000          0.0000
      1       0.3681       0.6323          0.3677
      2       0.1840       0.2642          0.7358
      3       0.0613       0.0802          0.9198
      4       0.0153       0.0189          0.9811
      5       0.0030       0.0036          0.9964
      6       0.0005       0.0006          0.9994
      7       0.0001       0.0001          0.9999

Probabilities less than .0001 have been omitted.

Where now:    E=EXPLORE menu     R=Run BINOM again     X=exit ? e
```

■ Exercises

7.1 Use BINOM to obtain tables for the following binomial probabilities:

a. n = 10, p = .1   f. n = 50, p = .5

b. n = 20, p = .1   g. n = 50, p = .3

c. n = 40, p = .1   h. n = 50, p = .1

d. n = 80, p = .1   i. n = 50, p = .02

e. n = 160, p = .1   j. n = 50, p = .005

7.2 For each of the BINOM p.d.f.'s in Exercise 7.1, describe the pattern of the probabilities. Sketch a rough graph (not an exact one) showing where the mean lies. Which of these p.d.f.'s would look acceptably like a bell-shaped curve, and which ones would not? Discuss the reasons.

7.3 Find the mean and the standard deviation of each of the binomial models in Exercise 1. Construct an approximate two-sigma interval whose end points are

$$\text{Upper end point:} \quad np + 2\sqrt{np(1-p)}.$$

$$\text{Lower end point:} \quad np - 2\sqrt{np(1-p)}.$$

Discuss the interval width in the various cases. In each case, does the lower end point include zero? Why would it matter? What happens to skewness, kurtosis, and the coefficient of variation as you go from Exercise 7.1a to 7.1e? From Exercise 7.1f to 7.1j?

7.4 In which of these binomial models would you trust a normal approximation? In which would you trust a Poisson approximation? Explain your reasoning fully.

7.5 Discuss what you think is the purpose behind the particular pattern of n's and p's shown in Exercise 7.1. What do you think can be learned from studying this particular sequence of models?

7.6 Experiment with "weird" binomials (by playing with n and p). See what you can learn about BINOM, and discuss your experimental findings. Did you cause any problems for BINOM by choosing "weird" experiments?

7.7 Use POIS to attempt Poisson approximations for the probabilities shown in Exercise 7.1. Use $\mu = np$ for each case. Are the tabulated probabilities from BINOM close to their equivalent tabulated probabilities from POIS? Would it matter for most purposes which method was used, based on these examples? In which cases are the tables most alike? What characteristics of n and p seem to be associated with a good equivalency between BINOM and POIS? Discuss your findings fully.

■ **References**

1. Hastings, N. A. J. and Peacock, J. B., *Statistical Distributions* (New York: John Wiley and Sons, 1975).

8

BOXPLOT/Box Plot for One Variable

Introduction

A *box plot* (or *box-and-whisker plot*) is simply a visual display of common statistics to reveal the general features of a data set. The end points (low and high data values) are always printed by BOXPLOT in the same position, with the *quartile points* displayed in a correct relative position along the horizontal axis. There will be a "box" in the middle, with "tails" on either side, extending to the end points. BOXPLOT also prints a few standard numerical statistics for reference purposes, perhaps obviating the need to run ANALYZ every time.

By looking at the box plot it is easy to see the width of the interquartile range in relation to the total data range and to tell whether the sample is symmetric. You can also see whether the median is placed at the midpoint of the box, thereby indicating symmetry *within* the middle range. Outliers are revealed by unusually long tails on the box plot. Box plots were conceived by John W. Tukey and are now a vital part of the field of *exploratory data analysis* (EDA for short).

The easiest way to appreciate a box plot is to look at a few. Figure 8.1 shows data that are skewed right, covering a relatively wide overall range (probably with outliers on the high end) but with the middlemost points packed into a rather narrow interquartile range.

Figure 8.2 shows a symmetric, bell-shaped sample with no outliers. While such data may not often be encountered in practice, it is a helpful prototype to keep in mind.

Figure 8.1   Box Plot Skewed to the Right

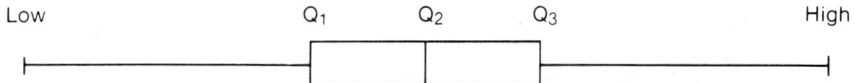

Figure 8.2   Box Plot of Symmetric Variable

Figure 8.3 shows a sample skewed somewhat to the left, with probable outliers at both ends. The tiny interquartile range suggests that many of the data values are packed into a rather small region in the middle.

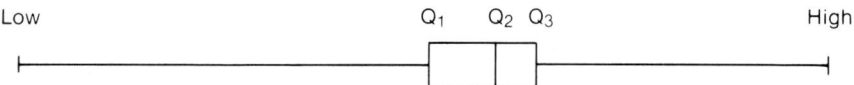

Figure 8.3   Box Plot Skewed to the Left

■ Case Study 8.1: Years Served by Supreme Court Justices

How long do U.S. Supreme Court justices serve? Table 8.1 has been compiled for all judges who retired between 1900 and 1982. The sample size is forty-five justices.

Using the EXPLORE editor, we create a file called JUDGES and then run the program BOXPLOT (see the printout on page 89). The range of years served is from 1 year (Byrnes) to 36 years (Douglas). Times center around a mean of 14.5 years. The median (second quartile) is only 13 years, so there is evidence that the mean has been pulled up by a few large data values. The mode is useless because there are eight values that each occur three times.

Which measure of central tendency gives the best notion of a "typical" judge's service? Both the mean and the median have merit. The mean is somewhat affected by the high values. The median, being insensitive to extremes, mitigates the influence of the three justices (Douglas, Black, Harlan) who served more than thirty years. Yet even these three justices were not extreme outliers, being within 2.5 standard deviations of the mean, so there is a good case for just using the mean.

The standard deviation is approximately 9.1, so the standard deviation is 63 percent as large as the mean. The implied coefficient of variation is about 63 percent. BOXPLOT uses the formula for a sample standard deviation, which seems appropriate in this case study.

Looking at the quartile points, we can say that about one-fourth of the justices served less than 6.5 years, while about one-fourth served more than 20.5 years. Rounding off a bit, it is approximately correct to say that the middle 50 percent of Supreme Court judges have served between 7 and 20

Table 8.1   Years Served by Supreme Court Justices

Name of Justice	Service Ended	Years Served	Name of Justice	Service Ended	Years Served
Stewart	1981	23	Brandeis	1939	22
Black	1971	34	Cardozo	1938	6
Douglas	1975	36	Sutherland	1938	15
Harlan	1971	16	VanDevanter	1937	26
Warren	1969	16	Holmes	1932	29
Fortas	1969	4	Sanford	1930	7
Clark	1967	18	Taft	1930	8
Goldberg	1965	3	McKenna	1925	26
Frankfurter	1962	23	Pitney	1922	10
Whittaker	1962	5	Day	1922	19
Burton	1958	13	Clarke	1922	5
Reed	1957	19	White	1921	10
Minton	1956	7	Lamar	1916	5
Jackson	1954	13	Lurton	1914	4
Vinson	1953	7	Harlan	1911	33
Rutledge	1949	6	Moody	1910	3
Murphy	1949	9	Fuller	1910	21
Stone	1946	4	Brewer	1910	20
Roberts	1945	15	Peckham	1909	13
Byrnes	1942	1	Brown	1906	15
Hughes	1941	11	Shiras	1903	10
McReynolds	1941	26	Gray	1902	20
Butler	1939	16			

Source: *The Hammond Almanac, 1982* (Maplewood, N.J.: Hammond Almanac, Inc., 1982), pp. 180–182.

years. U.S. Supreme Court justices have enjoyed considerable longevity, since one-fourth of them serve at least twenty years on the job.

Right-skewness is indicated by the fact that the mean exceeds the median. The sample skewness coefficient (0.577) also indicates right-skewness. Appendix F reveals that for a sample of size 50 (approximately our sample size) we would expect a skewness coefficient between −.534 and +.534 when sampling a normal population. Since our sample skewness coefficient is to the right of the upper end of the expected range, we would conclude that our sample probably did not come from a normal bell-shaped population.

The kurtosis coefficient (2.488) suggests a platykurtic (flatter than normal) sample. However, this statistic is consistent with normality, since the kurtosis coefficient lies within the range 2.15 to 3.99 specified in Appendix G for a sample of size 50.

We could interpolate Appendixes F and G if we wished (since n is 45 instead of 50), but the gain in accuracy would be slight. The sample's shape is best described as skewed to the right and slightly (but not significantly) platykurtic.

The box plot shows that a typical judge served fewer years than the mean, since the box is situated to the left of the center of the scale. Yet the median (Q_2) is fairly well centered between the first and third quartiles (Q_1 and Q_3). Although the data are skewed, the degree of skewness is not as strongly portrayed in the box plot as we might have anticipated. However, the data certainly are not symmetric.

In summary, the time served by a U.S. Supreme Court justice is typically about thirteen or fourteen years. About a quarter serve less than seven years, and about a quarter serve more than twenty years. The distribution is skewed

```
                          BOXPLOT

         This program will print a simple box-and-whisker plot
    (boxplot for short) plus a few common descriptive statistics
    for exploratory data analysis.  The boxplot will show the
    relative (visual) positions of the highest and lowest points,
    quartiles, and median.  You must have one disk file already
    saved, holding a column of up to 1000 observations.  You will
    be asked only to give the file name.

    File name? judges
    May I use graphics characters (y/n)? n


    S U M M A R Y    D E S C R I P T I V E    S T A T I S T I C S

    File  = JUDGES        Mean     = 14.48889    Skewness =   0.577
    Cases = 45            St. Dev. = 9.131918    Kurtosis =   2.488
    Low   = 1             1st Quartile = 6.5
    High  = 36            2nd Quartile = 13
    Range = 35            3rd Quartile = 20.5



            B O X P L O T    (File = JUDGES)


    LO            Q1            Q2            Q3                      HI
                  +------------+-------------+
    |----------|  |            |             |------------------------|
                  +------------+-------------+
```

Where now: E=EXPLORE menu   R=Run BOXPLOT again   X=exit ? e

to the right because a few judges have served an unusually long time, but there are no distinct outliers (at least since 1900). The distribution is not shaped like a normal distribution, being skewed to the right and slightly platykurtic.

■ Exercises

8.1 Choose any of the state data files from your DATABASE-1 diskette. Your sample size will then be n = 50. Run BOXPLOT, making a printout of the summary statistics and box plot.

8.2 Choose another state data file and run BOXPLOT. Obtain a printout of the summary statistics and box plot.

8.3 Compare these two box plots. How do the two data sets seem to differ in central tendency (width of box) and dispersion (width of box, separation from endpoints)? In making this comparison, rely on the box plot, not the numerical statistics. (You cannot compare the means, quartiles, or standard deviations because these units of measurement will differ—like comparing apples and oranges.) Does the purely visual approach tell you a lot about the data at a glance? What does it leave out, if anything?

8.4 Compare the shapes of the two data sets as implied by the skewness and kurtosis statistics. Use Appendixes F and G as required. Also, compute a variation coefficient. Write a succinct summary of your conclusions about shape.

8.5 Try to picture the histograms of these two data sets, based on the box plot and the other statistics. Then run ANALYZ to see what the histogram actually looks like. Were you able to anticipate the histogram accurately? If not, discuss why the data looked different than anticipated.

8.6 "In the future, statistics may become more visual and less mathematical. The box plot is an example. If we could also use color graphics, visual displays might replace most conventional statistical measurements." Discuss this statement fully.

Optional Exercises for Ambitious Learners

8.7 Run the program RAND to generate one hundred uniformly distributed random integers ranging from 0 to 1000. Save the resulting random numbers in a data file named UNIF-1. What would you expect the mean to be? The median? The quartile points? The standard deviation? The skewness coefficient? The kurtosis coefficient? Should this data set be symmetric? If you are not sure, answer as best you can. Then run BOXPLOT on the file UNIF-1. Check to see whether your predictions are fulfilled. Discuss the degree of departure of your sample from what you would have expected. Discuss the appearance of the box plot.

8.8 Repeat Exercise 8.7 using RAND to generate one hundred normally distributed random numbers with a mean of 50 and a standard deviation of 10. Save the resulting random numbers in a data file called NORM-1.

8.9 Repeat Exercise 8.7 using RAND to generate one hundred Poisson random integers with a mean of 9. Save the resulting random numbers in a data file called POIS-1.

■ References

1. Tukey, John W., *Exploratory Data Analysis* (Reading, Mass.: Addison-Wesley Publishing Company, 1977).

2. Hoaglin, David C., Mosteller, Frederick, and Tukey, John W., *Understanding Robust and Exploratory Data Analysis* (New York: John Wiley and Sons, 1983).

9

CHI/Chi-Square Test for Independence

Introduction

The program CHI will calculate a chi-square test statistic to measure the association between two variables that are cross-tabulated into a *contingency table*. This *chi-square test for independence* is a *nonparametric* test because no parameters are estimated. The only operation being performed is classifying the data into rows (variable one) and columns (variable two) and then comparing the *observed frequency* f(o) in each cell of the contingency table with the *expected frequency* f(e) under the assumption of independence.

If the two variables really are independent, then f(o) should be fairly close to f(e), resulting in a chi-square test statistic near zero. Large differences between f(o) and f(e) produce a chi-square test statistic that is considerably greater than zero, causing us to reject the hypothesis of independence. (Appendix D contains critical values of chi-square.) Degrees of freedom for the test are given by

$$d.f. = (r - 1)(c - 1)$$

where

r = the number of rows in the contingency table

c = the number of columns in the contingency table.

CHI has two options. The first option is to cross-tabulate a pair of data files that have been saved on the diskette and to perform a chi-square test for independence on the resulting contingency table. Each variable must be numerically coded, but the level of measurement need not be numerically meaningful. You will specify the size of the contingency table (number of rows, number of columns). The computer will offer advice about the proper size of table to attempt. The maximum is five rows and five columns.

Formation of classes is automatic unless you want to do it yourself. If left to itself, CHI will attempt to put about the same number of observations in each row and column. In forming classes, CHI will recognize the difference between continuous and integer data if the integers cover only a small range (such as 1, 2, 3, 4, 5) and will construct appropriate classes.

The second option is to have CHI calculate a chi-square test statistic from your data after they are already grouped into a contingency table. In this second option you need to type in only your observed cell frequencies, one row at a time. The rest of the test is automatic.

A table of observed frequencies will be printed, along with a table of expected frequencies (rounded to the nearest 0.1). If Yates's continuity correction is needed (when d.f. = 1), it will automatically be employed. If any cell frequency is below 5, a warning will be given, but calculations will proceed unless an actual mathematical error would result.

■ Case Study 9.1: Traffic Fatalities and Region

To study the relationship, if any, between state location and traffic fatality rates, the data shown in Table 9.1 may be used in a chi-square test for independence. The two data files TDEATH and WEST are part of the state database contained on the DATABASE-1 diskette. In file TDEATH, traffic fatality rates per 100,000 persons are used to remove the influence of population. A binary variable WEST is used to dichotomize the fifty states into West and East (1 = West, 0 = East), representing position relative to a line running approximately from Texas to Montana.

The hypotheses to be tested are

H_0:   Traffic fatality rates are independent of location of the state (West or East).

H_1:   Traffic fatality rates are *not* independent of location of the state (West or East).

Using the two files TDEATH and WEST, we run the program CHI (the printout is shown on pages 91–92). Comparing the summary table with the original data, we observe which states have the lowest traffic death rate (New York at 14.9) and the highest (New Mexico at 48.3). Although this comparison supports the hypothesis of a higher western traffic death rate, it means little because it compares only the extremes.

Accordingly, we look at the CHI printout showing data groupings into classes (three for traffic deaths, two for location). The computer automatically chooses the cutoff points for TDEATH. We might think of the three classes for fatality rates as low, medium, and high. Focusing on the extremes, we see that the first and third class limits are defined as follows:

Low = traffic fatality rate below 22.4 (sixteen states);

High = fatality of 29.2 or above (seventeen states).

Which states fall into which classes? Looking back at the original data (which may also be found in Appendix H), we make the tabulation shown in Table 9.2.

Table 9.1   Traffic Death Rate and State Location

State	Death Rate	West?	State	Death Rate	West?
Alabama	30.2	0	Montana	39.6	1
Alaska	20.4	1	Nebraska	22.4	0
Arizona	41.0	1	Nevada	45.2	1
Arkansas	32.3	0	New Hampshire	21.1	0
California	26.5	1	New Jersey	16.8	0
Colorado	25.3	1	New Mexico	48.3	1
Connecticut	18.8	0	New York	14.9	0
Delaware	19.9	0	North Carolina	27.6	0
Florida	28.4	0	North Dakota	21.3	0
Georgia	31.5	0	Ohio	21.4	0
Hawaii	23.3	1	Oklahoma	32.2	0
Idaho	39.3	1	Oregon	28.7	1
Illinois	19.9	0	Pennsylvania	19.7	0
Indiana	25.0	0	Rhode Island	15.8	0
Iowa	23.6	0	South Carolina	31.1	0
Kansas	24.4	0	South Dakota	32.9	0
Kentucky	26.6	0	Tennessee	29.1	0
Louisiana	30.5	0	Texas	31.0	1
Maine	24.3	0	Utah	24.7	1
Maryland	17.5	0	Vermont	30.4	0
Massachusetts	17.4	0	Virginia	19.3	0
Michigan	21.6	0	Washington	28.2	1
Minnesota	23.0	0	West Virginia	29.2	0
Mississippi	32.3	0	Wisconsin	21.9	0
Missouri	24.5	0	Wyoming	46.2	1

Source: *Accident Facts, 1983* (Chicago: National Safety Council, 1983), p. 63.

After studying Table 9.2, we begin to suspect that a variable called SOUTH might be a reasonable candidate for examination instead of WEST as a source of possible influence on TDEATH. However, the high-fatality states do include quite a few in the western United States, while the low-fatality states are mostly in the eastern United States.

Looking at CHI's table of observed frequencies (on the sample printout), we note that only one out of sixteen low-fatality states is in the West (column

Table 9.2   States Grouped by Traffic Fatality Rates

Low Fatality Rate:	Alaska, Connecticut, Delaware, Illinois, Maryland, Massachusetts, Michigan, New Hampshire, New Jersey, New York, North Dakota, Ohio, Pennsylvania, Rhode Island, Virginia, Wisconsin
High Fatality Rate:	Alabama, Arizona, Arkansas, Georgia, Idaho, Louisiana, Mississippi, Montana, Nevada, New Mexico, Oklahoma, South Carolina, South Dakota, Texas, Vermont, West Virginia, Wyoming

1 in the table of observed frequencies), while seven out of the seventeen high-fatality states are in the West (column 3 in the table of observed frequencies). From this preliminary analysis we expect to reject the hypothesis of independence when the chi-square test is performed.

CHI determines degrees of freedom from the number of rows and columns:

$$\text{d.f.} = (r - 1)(c - 1) = (2 - 1)(3 - 1) = 2.$$

From Appendix D we find the critical value to be 5.991, for the .05 level of significance. The decision rule for our hypothesis test is shown in Figure 9.1.

CHI's chi-square *test statistic* is found by using the formula

$$\chi^2 = \sum_{\text{all cells}} [f(o) - f(e)]^2/f(e).$$

Since CHI's test statistic (5.667) is not quite as large as the *critical value* (5.991), we cannot reject H_0. We cannot therefore conclude that traffic fatality rates are dependent on state location. However, if we were to use the .10 level of significance, the critical value (4.605) would cause us to reach the opposite conclusion.

We must conclude that this a borderline decision. There *does* seem to be some evidence of a relationship between traffic fatality rates and location. However, the relationship is not strong enough to be considered highly significant. The underlying causes need more investigation. Perhaps a regression model (using a program such as MGRES) would reveal a more precise idea of

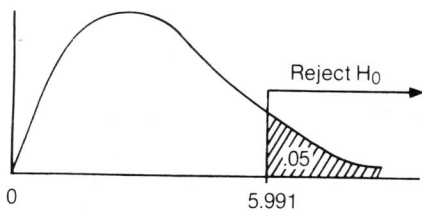

Figure 9.1   Decision Rule for Chi-Square Test

the factors associated with traffic fatality rates in the various states. Such factors could include the mean driving speed, alcohol consumption per capita, percentage of young drivers, and similar explanatory variables.

If the contingency table is 2 × 2 (that is, if d.f. = 1), CHI will subtract 0.5 from the absolute difference in the numerator of the test statistic before squaring. This correction is called *Yates's continuity correction.* Although Yates's correction is not relevant to our case study, it may apply to some of the exercises at the end of this chapter.

Another issue worth noting is whether our "sample" is perhaps really a population, since we are considering all 50 states. If so, a chi-square test really is unnecessary, since *any* departure from perfect independence would cause us to reject H_0. However, it could be argued that we are looking at these states at only one moment of time and that our data are only a sample of all possible observations on TDEATH that could be made. Whether this is semantics or not, the reader must decide.

```
                           CHI

          The chi-square test for independence of classification
     is based on a cross-tabulation of observed frequencies of two
     variables.  This cross-tabulation is also called a contingency
     table.  It may be constructed by reading two data files, or it
     may already be prepared.  You have two options:

          1. Raw data -- N observations on each of
             two variables, both saved in files

          2. Grouped data -- already tabulated into
             a contingency table of freqencies

     Which option do you want (1 or 2)? 1
     May I use graphics characters (y/n)? n


     SAVED DATA FILE INPUT:

     Program expects two saved data files of equal length.
     Maximum is 500 cases per file.  Computer will attempt
     automatic cross-tabulation if you wish, trying to choose
     class cutoff points to get approximately equal observed
     frequencies in row and column totals.

     Name of file #1: west
     Name of file #2: tdeath


          TABLE OF SUMMARY STATISTICS
     ==========================================
                     WEST           TDEATH
     ------------------------------------------
     Minimum         0              14.9
     Maximum         1              48.3
     Median          0              25.15
     Mean            .28            26.93
     Std. Dev.       .4535574       7.8403
     Sample Size     50             50
     ------------------------------------------
```

```
CHOICE OF CLASS LIMITS:

Want advice for preparing your cross-tabulation (y/n)? y
With 50 observations, your contingency table should
have no more than 10 individual cells if the expected
frequency is to be at least 5 in each cell.  Try to
choose the number of rows (R) and columns (C) so that
their product (RxC) does not exceed 10 .

For WEST: How many classes? 2
Want me to choose the class limits (y/n)? y

For TDEATH: How many classes? 3
Want me to choose the class limits (y/n)? y


GROUPINGS AND CLASS LIMTS FOR CONTINGENCY TABLE

For WEST:

Class          Value
-----          -----
  1              0
  2              1


For TDEATH:

Class          From           To (not incl)
-----          ----           -------------
  1            14.9           22.4
  2            22.4           29.2
  3            29.2           48.3
```

```
            TABLE OF OBSERVED FREQUENCIES

                    TDEATH
     WEST
                 1         2         3      Total:
             +---------+---------+---------+
       1     |   15    |   11    |   10    |   36
             +---------+---------+---------+
       2     |    1    |    6    |    7    |   14
             +---------+---------+---------+
     Total:       16        17        17      50


            TABLE OF EXPECTED FREQUENCIES

                    TDEATH
     WEST
                 1         2         3      Total:
             +---------+---------+---------+
       1     | 11.5    | 12.2    |   12.2|    36
             +---------+---------+---------+
       2     |  4.5    |  4.8    |   4.8 |    14
             +---------+---------+---------+
     Total:       16        17        17      50
```

```
Note: Some f(e) less than 5
```

Chi-Square statistic =   5.667 with d.f. = 2

```
Try again with different class limits (y/n)? n

Where now:  E=EXPLORE menu   R=Run CHI again   X=exit ? e
```

Case Study 9.2: Marijuana Views and Parent Dominance

A survey was given to a sample of fifty college students, asking the following questions:

1. "Looking at your own family, which response best describes your parents (use scale provided)?"

    Mother clearly   Father clearly
    dominant figure   1   2   3   4   5   dominant figure

2. "Should marijuana be legalized, subject to the same type of restrictions that apply to alcoholic beverages?"

    ____ No

    ____ Yes

A tabulation of the results is shown in Figure 9.2. The data seem ideally suited for a chi-square test to find out whether opinion on marijuana is independent of parent dominance.

Since the frequencies in some cells are too small to warrant using a 2×5 table (such a table may yield expected frequencies below 5 in the first and last columns especially), we "collapse" the table by combining responses 1 and 2 and responses 4 and 5:

Response 1 or 2   →   Mother dominant;

Response 3   →   Neither parent dominant;

Response 4 or 5   →   Father dominant.

The revised tabulation of responses using this new taxonomy is shown in Figure 9.3. Without any further calculations, several things are clear. First, the respondents oppose legalization of marijuana by a margin of about 2 to 1 (34 to 16, to be exact). Second, parent dominance seems about equally divided into the three categories, which might be expected. Third, it looks as if "father dominant" respondents are more opposed to legalization than "mother dominant" respondents.

Parent Dominance

Mother ←——————————→ Father

Legalize?	1	2	3	4	5	Row Total:
No	3	6	13	7	5	34
Yes	2	7	4	3	0	16
Column Total:	5	13	17	10	5	50

Figure 9.2 Tabulation of Marijuana Survey

Parent Dominance

Legalize?	Mother Dominant	Neither Dominant	Father Dominant	Row Total:
No	9	13	12	34
Yes	9	4	3	16
Column Total:	18	17	15	50

Figure 9.3   Revised Tabulation with Classes Combined

Using CHI, the observed frequencies shown in Figure 9.3 are entered row by row, separated by commas, as shown on the sample CHI printout following this case study. Short labels are used to help keep track of the meaning of CHI's rows and columns.

CHI's table of observed frequencies, of course, is identical to our hand tabulation. However, the table of expected frequencies does not appear to be drastically different from what would be expected if H_0 were true:

H_0: Opinion about marijuana legalization is independent of parent dominance.

H_1: Opinion about marijuana legalization is not independent of parental dominance.

CHI's degrees of freedom for the chi-square test are

$$d.f. = (r - 1)(c - 1) = (2 - 1)(3 - 1) = 2.$$

From Appendix D we find the critical value of chi-square for the .10 level of signficance to be 4.605.

Since CHI's chi-square test statistic (4.233) does not quite exceed the critical value (4.605), we cannot reject H_0. The effect of parent dominance, if any, is not very strong, since the .10 level of significance is not very stringent to begin with. (Type I error, or mistakenly rejecting a true null hypothesis, will be expected in about 1 in every 10 decisions at the .10 level of significance.) The decision rule is illustrated in Figure 9.4.

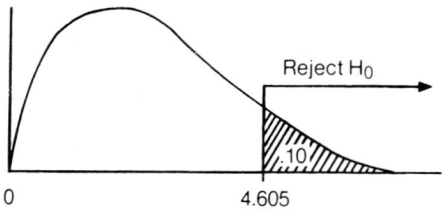

Figure 9.4   Chi-Square Decision Rule

Although the hypothesis of independence was not rejected, the effect of parent dominance might merit further study, since it was almost significant. Other demographic data (such as respondent's sex, age, religious beliefs, and so on) might also be considered as possible explanatory variables.

Note that one expected frequency is below 5. This is a slight violation of a commonly used rule of thumb but is not extreme enough to cause any problems. CHI does point out this small frequency, as a precaution.

```
                               CHI

        The chi-square test for independence of classification
is based on a cross-tabulation of observed frequencies of two
variables.  This cross-tabulation is also called a contingency
table.  It may be constructed by reading two data files, or it
may already be prepared.  You have two options:

        1. Raw data -- N observations on each of
           two variables, both saved in files

        2. Grouped data -- already tabulated into
           a contingency table of freqencies

Which option do you want (1 or 2)? 2
May I use graphics characters (y/n)? n



GROUPED DATA INPUT OPTION

Would you like to assign SHORT descriptive names
to each of your variables (y/n)? y
Name of first variable (columns)? Dominant
Name of second variable (rows)? Legalize

Maximum size is 5x5 (5 rows, 5 columns).
How many rows? 2
How many columns? 3

When requested, input the observed cell frequencies for
each complete row, going across.  Use commas to separate
frquencies (example: 12,21,22).  No comma at end of line.
Input 3 cell frequencies for row 1 ? 9,13,12
Input 3 cell frequencies for row 2 ? 9,4,3



            TABLE OF OBSERVED FREQUENCIES

                   Dominant
      Legalize
                     1         2         3      Total:
                 +---------+---------+---------+
           1     |    9    |   13    |   12    |     34
                 +---------+---------+---------+
           2     |    9    |    4    |    3    |     16
                 +---------+---------+---------+
      Total:         18        17        15          50
```

```
           TABLE OF EXPECTED FREQUENCIES

                    Dominant
   Legalize
                    1           2           3        Total:
                +---------+---------+---------+
        1       |   12.2  |   11.6  |   10.2  |        34
                +---------+---------+---------+
        2       |    5.8  |    5.4  |    4.8  |        16
                +---------+---------+---------+
   Total:           18          17          15         50
```

Note: Some f(e) less than 5

Chi-Square statistic =   4.233   with d.f. =   2

Where now: E=EXPLORE menu   R=Run CHI again   X=exit ? e

■ Exercises

9.1 In a certain corporation, employees are asked to complete a QWL (Quality of Work Life) survey. Their responses are shown below. Use CHI's second option (grouped data) to enter the frequencies from this contingency table, and obtain a printout of the expected frequencies and chi-square test statistic. At the .05 level of significance, is there any difference between the hourly and salaried workers' job satisfaction? Verify the degrees of freedom. Illustrate your decision rule. Is the decision sensitive to the level of significance chosen? Are the expected frequencies in all cells large enough to ensure a valid test? Discuss fully.

	Satisfied	Neutral	Not Satisfied	Total:
Salaried	20	13	2	35
Hourly	135	127	58	320
Total:	155	140	60	355

9.2 In a certain county, recipients of public assistance were asked to complete a questionnaire to ascertain the length of time required to process their claims prior to receipt of payments. Their responses are shown below. Use CHI's second option (grouped data) to enter the frequencies from this contingency table and obtain a printout of the expected frequencies and chi-square test statistic. At the .05 level of significance, is there any relationship between the applicant's ethnicity and the delay encountered? Verify the degrees of freedom. Illustrate your decision rule. Is the decision sensitive to the level of significance chosen? Are the expected frequencies in all cells large enough to ensure a valid test? Discuss fully.

	No Delay	Some Delay	Long Delay	Total:
White	73	62	60	195
Black	25	31	39	95
Hispanic	24	26	35	85
Other	5	4	11	20
Total:	127	123	145	395

9.3   Choose any two state data files from the DATABASE-1 diskette that you think might be related. Use CHI's first option (raw data) to obtain a cross-tabulation of these two files, using any number of rows and columns (up to 5 × 5). Record the chi-square test statistic and degrees of freedom. Then experiment with other table sizes, varying the number of rows and/or columns. Be careful to note if the expected frequencies get too small (problems begin to occur only if any expected frequency drops *far* below 5). Reduce the number of rows or columns if necessary to be sure that at least some of your tables have large enough expected frequencies.

Now set up a decision rule using the .10 level of significance, and test each CHI table for independence. Illustrate your decision rule. What would happen if you used the .05 level of significance instead? The .01 level of significance? The .005 level of significance?

Discuss whether your decision is sensitive to the level of significance. Also, discuss whether your decision is sensitive to the particular cross-tabulation chosen. That is, does varying the number of rows and columns affect your decision? Discuss fully.

9.4   A study of vitamin C was carried out in Toronto, Ontario, in which 407 subjects received one gram daily and 411 subjects received none. Their susceptibility to the common cold was then studied. This was a double-blind experiment in which neither experimenters nor subjects knew whether vitamin C had been administered. Carry out a chi-square test on the results shown below to see whether the proportion of colds differs significantly between the control and experimental groups. Discuss your conclusions fully. State the hypotheses clearly. Is the continuity correction needed? What else do you see in the tabulation of data that should be discussed?

	Received No Vitamin C	Received Vitamin C	Total:
Got a Cold	337	301	638
Did Not Get a Cold	74	106	180
Total:	411	407	818

Source: "Vitamin C: On the Defensive Again," *Science News* 107 (March 22, 1975): 191.

9.5   Sixty-four students in introductory college economics were asked to state how many credits they had earned in college and to indicate how certain they were about their choices of major. Carry out a chi-square test on the results shown below to see whether student certainty is independent of the number of credits earned. Discuss your conclusions fully. State the hypotheses clearly. Is the continuity correction needed? Do you need to worry about expected frequencies being too small? If so, is there anything to be done about it? What else do you see in the tabulation of data that should be discussed?

	Not Very Certain	Fairly Certain	Absolutely Certain	Total:
Under 10 Credits	12	8	3	23
10 to 59 Credits	8	4	10	22
60 or More Credits	1	7	11	19
Total:	21	19	24	64

Source: Classroom survey.

9.6   A large state university published a map showing how many alumni lived in each county in the state and the number of currently enrolled students from each county. The frequencies are shown below for two randomly chosen counties. Choose a level of significance and carry out a chi-square test to see whether the observed differences in alumni and student proportions are significant (in other words, whether the student/alumni ratio is independent of county). Discuss your analysis fully, including the continuity correction, small expected frequencies (if any), and an eyeball assessment of the table of observed results. [Optional: If you can guess which state university it is, you get extra credit.]

	Morris County	Russell County	Total:
Alumni	50	111	161
Students	12	50	62
Total:	62	161	223

9.7   The table below shows the number of people killed or injured by lightning in the United States in two years, 1950 and 1960. Choose a level of significance and carry out a chi-square test using CHI to see whether the proportions of killed and injured are significantly different in the two years (that is, whether severity of accident is independent of year). Discuss your analysis fully, including the continuity correction, small expected frequencies (if any), and an eyeball assessment of the table observed results.

	Killed	Injured	Total:
1950	59	78	137
1960	100	199	299
Total:	159	277	436

Source: Martin A. Uman, *Understanding Lightning* (Carnegie, Penn.: Bek Technical Publications, 1971), p. 19.

9.8 A university survey of randomly chosen new students revealed the data below. Discuss what the table tells you *prior* to running any formal tests. What would a decision maker miss by not breaking the table down into rows but only looking at the column totals? Set up appropriate hypotheses and test them, using CHI for the calculations. Explain fully, illustrate your decision rule clearly, specify your level of significance, and discuss anything else of importance.

	Low Tuition	Good Location	Academic Reputation	Total:
Freshmen Students	50	30	35	115
Transfer Students	15	29	20	64
Graduate Students	5	20	60	85
Total:	70	79	115	264

Optional Exercises for Ambitious Learners

9.9 Repeat the analysis in Exercise 9.8 using only the 2 × 3 comparison between the first two rows (Freshman Students and Transfer Students). Discuss your results in comparison with the previous test on the whole table. Why might a decision maker want to look only at the first two rows?

9.10 In each of the 2 × 2 tables (see Exercises 9.4, 9.6, and 9.7), carry out a z test for two proportions, using the program TWOSAM. See whether you can reject the hypothesis of equal proportions:

$$H_0: \quad p_1 = p_2.$$
$$H_1: \quad p_1 \neq p_2.$$

For example, in Exercise 9.4, use the following sample estimates for the proportion of participants who did not get a cold:

$$\hat{p}_1 = 74/411 = .1800 \quad \text{and} \quad \hat{p}_2 = 106/407 = .2579.$$

How do the results of this test compare with the chi-square? Do you see evidence that the z test is more powerful? Discuss fully, showing all your reasoning clearly.

9.11 Design a simple questionnaire (two to five questions only). One question should be the object of your survey, and the other(s) should be factors you want to control (such as respondent's gender). Avoid excessively nosy questions that people will not want to answer (such as income or sex habits) unless you can guarantee anonymity. Worry about the wording of your questions, and choose scientific scales that are easily understood by respondents. State your hypotheses clearly.

Check first with whoever is in charge of the place where you take your survey to make sure that it is allowed. It is best to choose a known test group, such as classmates, or co-workers so that your survey does not become too intrusive. When you have finished your random sample of respondents, set up your hypotheses about independence between two of your questions. Then run CHI, using the first option (raw data) if you put your raw data in disk data files or the second option (grouped data) if you have already tabulated the results.

Discuss your sample results. What do you think is the level of accuracy in the responses? What is the level of measurement of the data? What problems might exist in the data collection? What should have been done differently?

Discuss your cross-tabulation thoroughly, trying to squeeze out any implications that it might hold, before doing a formal chi-square test. Combine classes if necessary. Carry out the chi-square test, choosing any level of signficance. Verify the degrees of freedom. Illustrate your decision rule(s). Experiment with different data classifications if you think it is appropriate (discuss why or why not). Discuss the implications of your decision. Is it sensitive to the choice of level of significance? To the way data classes are chosen? Are expected frequencies large enough? Was a continuity correction needed? How helpful was the computer?

10

DICE/Law of Large Numbers Using Dice

Introduction

Much of the early research in probability and statistics was motivated by the study of common gaming devices such as cards and dice. A statistics student can still learn a lot from a careful study of ordinary dice. The DICE program simply turns the computer into a rapid and tireless dice player who keeps careful records of everything that happens. DICE is designed to help you understand an important principle known as the *Law of Large Numbers*.

The Law of Large Numbers says, among other things, that as the sample size increases, what *actually* happens will get relatively closer to what *theoretically* should happen. In small samples there is much less tendency for this to occur. For example, if you flipped a fair coin one hundred times, you would be surprised if you got seventy-five heads. Yet if you flipped this same coin four times and got three heads, it would not seem worth mentioning. Departures from the expected value become dramatic and convincing only if the sample size is large. If the departure is significant, you would doubt the fairness of the experiment. For example, if you really got seventy-five heads in one hundred tries, the coin's owner would be in for some tough questions.

Everyone who has seen James Bond win at dice can appreciate that winning can be impressive when the stakes are high. But anyone who has spent a lot of time playing dice in Las Vegas or Atlantic City knows that it is impossible to win time after time, day after day. Although gamblers may want to believe otherwise, dice have no memory. This fact does not prevent an individual from having a lucky streak, though. Therein lies the source of people's interest in games (if nobody ever won, gambling's popularity would have faded long ago). The odds against ''long streaks'' are low, but some impressive ones have been witnessed.

Unfortunately, gambling is an expensive (and possibly addictive) method of exploring the Law of Large Numbers. As an alternative, you could flip coins or roll dice on your own and carefully tabulate the results. But this could become time-consuming, and the object of study could be lost in the tedium of data collection. The program DICE can simulate in a few minutes experiments that would take you hours to carry out on your own. By studying the outcomes you can see how close your experiments come to theoretical predictions. There is a regular simulation option and a game option. The game is intended to focus your attention on the *relative* difference between observed and expected frequencies, as you vary the number of rolls and the number of dice.

■ Case Study 10.1: Rolling One, Two, and Three Dice

A *fair die* is a small cube of uniform density with smooth sides. It is reasonable to assume that when such a die is bounced on a table or floor, each side (face) is equally likely to be on top when the cube comes to rest. Since there are six faces (labeled with 1, 2, 3, 4, 5, or 6 dots), the probability of any particular number's appearing on a single throw is 1/6, and the expected number of appearances in n throws is n/6. For example, if we throw a fair die thirty times, the expected number of 3s is five. However, the probability that 3 will appear *exactly* five times is rather small. This fact is a paradox to many people.

To generalize, let X represent the number of appearances of a particular face in n throws of a fair die. X will be a random variable that can assume any value from 0 through n. The probability of exactly r appearances of a particular face is the binomial probability

$$P(X = r) = \frac{n!}{r!(n - r)!} (1/6)^r (1 - 1/6)^{n - r}$$

where

n = number of times the die is thrown

p = probability of a particular outcome on each throw = 1/6.

Here, we define *success* as the appearance of the face we are looking for. For example, the probability of exactly five 3s in thirty throws is

$$P(X = 3) = \frac{30!}{5! \, 25!} (1/6)^5 (1 - 1/6)^{25}$$

$$= .1917.$$

With one die, all outcomes have the same probability; with two or three dice, things are a little more complicated. When two dice are thrown, the outcome of interest is the number of dots on the two faces that turn up:

$$X = X_1 + X_2$$

where

X = total showing on both dice

X_1 = number showing on first die

X_2 = number showing on second die.

	Outcome of First Die (X_1)					
	1	2	3	4	5	6
1	2	3	4	5	6	7
2	3	4	5	6	7	8
3	4	5	6	7	8	9
4	5	6	7	8	9	10
5	6	7	8	9	10	11
6	7	8	9	10	11	12

Outcome of Second Die (X_2)

Figure 10.1   Sample Space for Sum of Two Dice

The thirty-six possible outcomes are shown in Figure 10.1. We can see that there are four ways of getting a 9. Since each of the thirty-six possible sums is equally likely, the probability is 4/36 that we will see 9 showing on the dice. We are intuitively familiar with most of these facts from playing games (such as Monopoly). For example, 7 is the most likely outcome, and 2 or 12 occur only rarely (once in thirty-six tries).

The binomial probability model can be applied to any two-dice experiment, using Figure 10.1. For example, the probability of getting exactly twelve 8s in seventy-two throws is

$$P(X = 12) = \frac{72!}{12!\ 60!}\ (5/36)^{12}\ (1 - 5/36)^{60}$$

$$= .1005 \quad \text{or about 10 percent.}$$

Clearly, this calculation would be difficult without some kind of computer program. The result above was obtained using BINOM with parameters:

$$n = 72;$$
$$p = .1389 \quad \text{(since } 5/36 = .1389\text{)}.$$

With BINOM you can also find sums of probabilities in either tail of the binomial p.d.f. For example, from BINOM the probability of getting fewer than twelve 8s in seventy-two throws is

$$P(X < 12) = .7057 \quad \text{or about 71 percent.}$$

The case of rolling three dice is similar to that of two dice, except that the sample space is more complicated. For two dice there are 36 possible outcomes, while for three dice there are 216 possibilities! All the possible outcomes cannot be listed here, but their probabilities are summarized for reference purposes in Table 10.1.

DICE will roll one, two, or three dice. The number of rolls is up to you. You will see each roll, plus a histogram to summarize the outcomes. For each

Table 10.1. **Summary of Dice Probabilities**

One Die		Two Dice		Three Dice	
Dots	Probability	Dots	Probability	Dots	Probability
1	.1667 (1/6)				
2	.1667 (1/6)	2	.0278 (1/36)		
3	.1667 (1/6)	3	.0556 (2/36)	3	.0046 (1/216)
4	.1667 (1/6)	4	.0833 (3/36)	4	.0139 (3/216)
5	.1667 (1/6)	5	.1111 (4/36)	5	.0278 (6/216)
6	.1667 (1/6)	6	.1389 (5/36)	6	.0463 (10/216)
		7	.1667 (6/36)	7	.0694 (15/216)
		8	.1389 (5/36)	8	.0972 (21/216)
		9	.1111 (4/36)	9	.1158 (25/216)
		10	.0833 (3/36)	10	.1250 (27/216)
		11	.0556 (2/36)	11	.1250 (27/216)
		12	.0278 (1/36)	12	.1158 (25/216)
				13	.0972 (21/216)
				14	.0694 (15/216)
				15	.0463 (10/216)
				16	.0278 (6/216)
				17	.0139 (3/216)
				18	.0046 (1/216)

experiment you will be given actual and expected frequencies for each possible outcome. This notation is used:

f(o) = actual number of times the outcome occurred;

f(e) = expected number of times outcome should occur.

You should look for instances in which there is a large difference between f(o) and f(e). When you spot outcomes for which you think the difference between f(o) and f(e) is beyond the level you would attribute to random chance, you may check the probability of events like P[X ≥ f(o)] or P [X < f(o)] using BINOM with

n = number of times you rolled the dice

p = probability of each roll (from Table 10.1).

In other words, Table 10.1 will help you set up BINOM runs to check the probabilities of each of your DICE experiments if you wish.

DICE also offers you the option of playing a game, designed to focus attention on the magnitude of differences between actual and expected outcome frequencies. For example, when you roll one die thirty-six times, you "should" get six 4s. Yet you would not be surprised if 4 appeared five or seven times instead of exactly six times.

How close would you expect to come to the expected value? Within 20 percent? Within 50 percent? Within 100 percent? The Law of Large Numbers would lead you to expect that as you roll the dice more often, the actual outcome frequencies should approach their expected values:

$$\lim_{n \to \infty} \frac{f(o)}{f(e)} = 1.$$

Percent differences should become smaller as n becomes larger.

If you specify an allowable percentage error (call it A), you could use the number of times f(o) and f(e) differ by A or more (call it C) as a measure of how far apart actuality is from theory. The game assigns you a score (S), using this formula:

$$S = \log(1 + nA) + 2C.$$

The formula is probably not worth trying to figure out (that is not the main point anyway). All you need to know are six facts:

Fact #1:   Low scores are better than high scores.

Fact #2:   Your score goes up when you increase any of the variables (n, A, or C).

Fact #3:   You may set n and A yourself.

Fact #4:   C is determined by n and A.

Fact #5:   C tends to drop when you increase n.

Fact #6:   C tends to drop when you increase A.

As in real life, there are tradeoffs. For example, you can reduce C by raising n, but raising n increases your score. You can reduce C by raising A, but raising A increases your score. "That's not fair," you say? Say instead, "Here's a challenge!" After all, a game should be challenging, or it would be no fun. To reduce your score, you can just use trial-and-error.

Each number of dice (one, two, three) will be considered a different game. A summary of your scores will be printed when you are finished. DICE will offer advice if it senses that you are "spinning your wheels." If you play the game, a little patience and experimentation will go a long way.

How does DICE simulate the behavior of small wooden or plastic cubes? Without a computer there are many ways. All we require is that each of six outcomes be equally likely. For example, we could spin a wheel of fortune whose circumference is divided into six equal parts. Another way is to choose one-digit random numbers from a table, discarding 0, 7, 8, and 9.

The computer's method is a modified version of the random-digit method. Using the BASIC random number function RND, we choose a random decimal number between 0 and 1 and multiply it by six. Then we add 1 and truncate it to the next lower integer to yield X, the simulated outcome of the die's roll:

$$X = INT(1 + 6*RND)$$

This statement is placed in a loop to generate n simulated rolls of one die. Two or three dice, of course, are just the sum of separate rolls of one single die.

The following printout illustrates what happens when we roll one die 120 times, two dice 72 times, and three dice 216 times. The printout is self-explanatory. Notice that the outcomes will depend on the initial random starting point that you specify (a one-digit number). The game option is not illustrated with a sample printout, partly because it looks about the same and partly because it is going to depend substantially on your individual decisions. The exercises at the end suggest some ways of using both the regular simulation and the game to explore the behavior of DICE systematically.

A final caution: Even using the same random "seed," you might not be able to duplicate the exact printout shown here, although the format will appear identical. It all depends on how your computer's random number sequence is arranged and initialized.

```
                          DICE

        In this random experiment I will simulate the
rolling of 1,2, or 3 dice. You will tell me how many
times to roll. Then I will print a histogram of the out-
comes. I will also show values for f(o) and f(e), where:

        f(o) = the actual frequency of throws of each type
        f(e) = the expected theoretical frequency of throws

If you wish, the dice rolls can be made into a mathematical
game, based upon a comparison of f(o) with f(e).

Shall I explain the mathematical game (y/n)? n
Do you want a game (1) or a regular simulation (2)? 2
For random start, input a 1-digit number? 3
How many dice shall I throw (1,2, or 3; 0=stop)? 1
How many times shall I throw? 120


                   120 ROLLS OF 1 DIE:

      4   6   1   4   2   2   2   6   6   4   3   2   2   1   6   1   6   6   2
      3   3   2   5   4   1   2   2   5   3   6   5   6   5   6   6   5   5   3
      2   4   5   5   3   1   6   4   1   5   1   3   6   6   1   6   5   4   1
      4   3   4   1   2   4   4   1   1   5   2   6   5   1   4   1   3   5   5
      3   6   5   2   5   6   6   1   6   5   1   3   4   3   6   2   5   4   1
      2   1   1   5   4   2   4   1   1   6   6   1   1   2   6   3   1   5   6
      4   6   3   1   3   5
```

```
         EXPECTED   ACTUAL
  DOTS     FREQ      FREQ    HISTOGRAM FOR ONE DIE
  ------------------------------------------------------------
    1        20        25    *************************
    2        20        17    *****************
    3        20        15    ***************
    4        20        17    *****************
    5        20        21    *********************
    6        20        25    *************************
  ------------------------------------------------------------
How many dice shall I throw (1,2, or 3; 0=stop)? 2
How many times shall I throw? 72
```

```
                    72 ROLLS OF 2 DICE:

    7   6   7   9   8   6   9  10   7  10   8   8   6   7   7   6   8
    5   6   7  11   8   8  10  11   7   8   9   5   9   7   6  11   7
    8   4   6   6   5   8   6   6   2   8  10   6   6   6   8   7   9
   10   4   7   8   9  10   4   7  10   7   8   7   7   3   5   8   7
    7   2   9   7
```

```
         EXPECTED   ACTUAL
  DOTS     FREQ      FREQ    HISTOGRAM FOR TWO DICE
  ------------------------------------------------------------
    2         2         2    **
    3         4         1    *
    4         6         3    ***
    5         8         4    ****
    6        10        13    *************
    7        12        18    ******************
    8        10        14    **************
    9         8         7    *******
   10         6         7    *******
   11         4         3    ***
   12         2         0
  ------------------------------------------------------------
How many dice shall I throw (1,2, or 3; 0=stop)? 3
How many times shall I throw? 216
```

```
                    216 ROLLS OF 3 DICE:

   10  17   6  10   7  12   7   9  16   9  15  13  13  15   7
   12   3  10  14   6  10  10  11  13  13  11  11   5   7  13
    8   8  10   6  10   4  13  11  11  12   8  12   8  15   9
   12   8  12  11  13  13  13   8   8  17   8   9   6   9   8
   11  11  11  11  14  11  14  11  10   8  13  17  12   6
   14  10  15   8  12  12  18  10   7  10  11   7  10  15  14
    6   8   9  11  11   7   5  13   9  12  12   8  10   4  11   5  18
   10   9  11  11   8  11  10  10  11  11   7  12  12   4   8
    6  12  12   3   9   9  14  10   5   8   9  12   5  10  11
   10  10   5  11  13  10  10  10  17   8  10  12  12  10
   11  12  10  12  10  14   9   8   9  12  12  15   9   7  14
   13  10   8   6   8  11   9   8  10  14  11   8  15   7  12   5
   16   8  10  12  11   9   9  11  16  10   8  13  16  11  17
   14   8  17  10  12  10  10  14  14   9   9  10  11  10
   10  14   8  11
```

```
        EXPECTED   ACTUAL
  DOTS    FREQ      FREQ    HISTOGRAM FOR THREE DICE
  --------------------------------------------------------------------
    3       1         2     **
    4       3         3     ***
    5       6         7     *******
    6      10         8     ********
    7      15        10     **********
    8      21        26     ************************
    9      25        19     *******************
   10      27        38     **************************************
   11      27        31     *******************************
   12      25        25     *************************
   13      21        15     ***************
   14      15        13     *************
   15      10         7     *******
   16       6         4     ****
   17       3         6     ******
   18       1         2     **
  --------------------------------------------------------------------
How many dice shall I throw (1,2, or 3; 0=stop)? 0

Where now:   E=EXPLORE menu    R=Run DICE again    X=exit ? e
```

■ Exercises

Regular Simulation Option

10.1 Experiment with one die. Start with a small number of rolls, and increase the number of rolls until the shape of the resulting distribution is clear to you. Note that if the number of rolls is not an even multiple of 6, the expected frequencies will not be integers, and cannot equal the observed frequencies. Print the histogram if you can.

10.2 Roll one die 216 times. Roll two dice 216 times. Roll three dice 216 times. If the shapes of the distributions are not clear of if you suspect that you got an unusual result, try it again. Print the histograms if you can.

10.3 Try a really large n on at least one experiment of your choice. The limit is n = 1000, but that many rolls will take awhile, so be patient if you go that high. Print the histogram if you can.

10.4 Discuss the general appearance of the sample distributions you obtained for one die. Calculate the ratio f(o)/f(e) for each of your experiments, and plot each ratio on some kind of histogram-type display using a scale like this:

f(o)/f(e) Ratios for One Die with n = 60

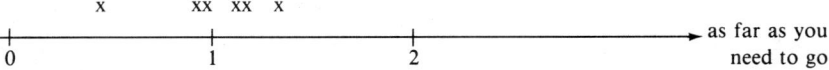

Each display should have six x's. There should be one display for each sample size. Discuss what your displays tell you about the Law of Large Numbers.

10.5 Compare the shapes of the histograms for one, two, and three dice in the n = 216 samples taken in Exercise 10.2. Can you make any generalizations about central tendency and dispersion as the number of dice increases (holding n constant at 216)? What prediction would you make about the appearance of the histogram for four dice rolled 216 times?

10.6 Run BINOM to obtain (and print if possible) binomial probability tables for the following cases:

a. Your smallest n with one die, using p = .1667.

b. Your largest n with one die, using p = .1667.

c. Any other experiments you want to try.

d. n = 60 and p = .1667 (if not included in 10.6a–10.6c).

10.7 Study your binomial probabilities from BINOM. Circle the ones that are relevant to your most unusual outcomes [that is, cases in which f(o) and f(e) differ greatly] and write down the probabilities that the outcomes would have been less extreme. Write a summary of your findings about the probability that such things would have happened by chance. How do you explain what actually happened?

10.8 A student complained, "I threw one die sixty times. I should have obtained ten 6s. But I only got eight 6s, so I know the die was biased. What a cheat! The computer doesn't play fair!" Is this complaint justified? What is the probability of obtaining exactly ten 6s in sixty rolls of one fair die? Refer to Exercise 10.6d in answering.

10.9 As a slightly harder exercise for the more ambitious learner, run BINOM to obtain a probability table for n and p for your most extreme difference between f(o) and f(e) with one die, using Table 10.1 to obtain p. Repeat for two dice. Repeat for three dice. Discuss your findings fully.

Game Option

10.10 Since the object of the exercise is to get the smallest scores you can, no real instructions are needed. Just start playing, and quit when you are satisfied. The following suggestions may help you get started:

a. Begin with one die. Hold your allowable error (A) constant and increase or decrease n to see how much effect sample size alone has.

b. Now hold the number of rolls constant at some fairly high level (in other words, avoid tiny sample size) and try changing the allowable error (A) until you get a feeling for how much effect A alone has.

c. Move on to two dice, then three dice. Use a systematic strategy such as those mentioned in Exercise 10.10a or 10.10b.

10.11 When you are finished, make a copy of your scores, and then write an analysis of what happened as you played the game. What worked? What didn't? How would you advise someone else to win this game? If you had more time, could you improve on your own scores? Why, or why not? How precise do you think the Law of Large Numbers is? Explain.

11 GOODFT/
Goodness-of-Fit Tests

Introduction

GOODFT performs a chi-square *goodness-of-fit test* to see whether your sample resembles a particular kind of population. GOODFT also prints a few common statistics and visual displays. Often, a simple eyeball inspection of the data or statistics will reveal as much as a goodness-of-fit test. For example, is the sample strongly bimodal or skewed? Are outliers present?

Tests for goodness of fit are easy to understand in a general way. But in the old days the tedious calculations tended to discourage their use. The computer has drastically changed the situation. Today, goodness-of-fit tests are often easier to perform than other hypothesis tests because the computer makes the calculations almost automatic.

GOODFT offers automatic goodness-of-fit tests for three different models: normal, uniform, and Poisson. The three models correspond to the random-number-generating capabilities offered by the program RAND. There is also a "do-it-yourself option" to aid you in working with other situations (such as binomial model). The hypotheses to be tested are of the form:

H_0:   The population of X's follows a _____ model.

H_1:   The population of X's does not follow a _____ model.

The blank may contain the name of any theoretical model (such as normal, uniform, Poisson). The choice of model depends on your situation. Using data grouped into k classes (by you or by the computer), GOODFT finds the chi-square test statistic

$$\chi^2 = \sum_{\text{all } k} \frac{[f(o) - f(e)]^2}{f(e)}$$

where

f(o) = observed class frequency in your sample

f(e) = expected class frequency assuming H_0 is true.

If f(o) and f(e) differ greatly, the test statistic will be large. We will reject H_0 if the test statistic exceeds the critical value chosen from Appendix D. Degrees of freedom are

Normal:	d.f. = k − 3	(2 parameters estimated)
Poisson:	d.f. = k − 2	(1 parameter estimated)
Uniform:	d.f. = k − 1	(0 parameters estimated).

Of course, your sample may not resemble any of these models unless there is a single, constant underlying cause.

Normal Model

The normal model would be appropriate for any continuous variable with a reasonable degree of central tendency unless serious skewness is present (usually caused by outliers). Scientific measurements of time or physical attributes (weight, size, distance) are a few obvious possibilities. The normal model also applies to discrete data if the range is relatively large, such as test scores or number of occurrences per unit of time (if the mean is large). GOODFT is ideal for analyzing random normal samples generated by program RAND.

Any normal population is fully described by two parameters: the mean and the standard deviation. From a file containing the ungrouped sample data, GOODFT will estimate these two parameters. Then each X is transformed into a standardized z value, using

$$z = \frac{X - \text{Sample Mean}}{\text{Sample S.D.}}.$$

We know that for a normal distribution the relative frequency of observed z values in the sample would be as shown in Figure 11.1.

GOODFT simply multiplies these areas by the sample size (n) to obtain expected frequency for each class. In the end classes, very few observations would be expected. Since small expected frequencies can cause trouble for a chi-square test, GOODFT automatically collapses classes from the ends in-

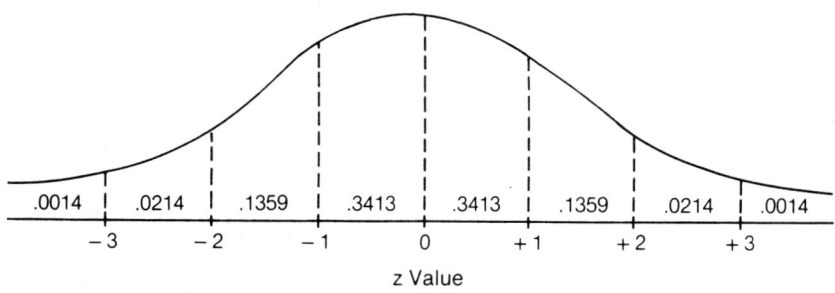

Figure 11.1 Normal Probability Density Function

ward, as needed, to enlarge the expected frequencies. Therefore you will actually have fewer than eight classes in the goodness-of-fit test.

After classes have been collapsed as much as possible, a warning is printed if any expected frequency remains below 5 (a widely accepted rule of thumb). If any expected frequency is below 2, the calculation halts. This condition arises if your sample size is very small. With a small sample you really cannot tell whether the population is bell-shaped. As a guideline, a normal goodness-of-fit test should be avoided if n < 25.

Uniform Model

Uniform data are rare in daily life. Some possibly uniform patterns occur over days of the week (such as volume of stock traded or phone calls received in a business) or over months of the year (such as insurance claims filed or drunks arrested). Other things are "supposed" to be uniform, such as lottery winning numbers over a series of weeks or random numbers generated by a computer (such as the program RAND). Often, we do not really believe that uniformity exists, but we use it as a "straw man" to be knocked down in order to prove that something is not uniform.

The computer will automatically read your sample of ungrouped observations from a file (such as the one saved by RAND) and will set up k classes of equal width. Assuming that the sample data are uniformly distributed, the expected frequency in each class should be n/k because every class is equally likely (see Figure 11.2). GOODFT automatically sets the number of classes, class limits, and interval widths in such a way that the resulting classification will (hopefully) be aesthetically pleasing.

If your data are already grouped into k classes, you may input the observed frequencies directly, without asking GOODFT to read a file. This option adds flexibility, just in case you want to experiment on your own.

In either case, GOODFT calculates the chi-square test statistic in the same way. If your sample size is small, the expected frequencies in each class could become too small, which might cause difficulty. A warning is printed if any expected frequency is below 5, and the calculation is terminated if any ex-

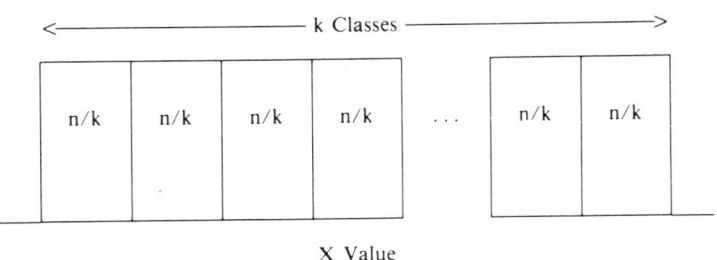

Figure 11.2 Uniform Probability Density Function

pected frequency is below 2. (Since all expected frequencies are the same, this condition will apply to all classes.)

Poisson Model

In a Poisson model, X must represent the number of *events per unit of time (or space)*. Events that tend to fit this definition include points scored in sports events (such as number of touchdowns per football game, number of goals scored per soccer game, number of home runs per baseball game). Other standard applications include customer arrivals per minute in a bank or supermarket, information calls per minute at a switchboard, alarms per hour at a fire station, computer breakdowns per day, and so on. The program RAND will generate Poisson data and will save the numbers in a file that is ideally suited for analysis by GOODFT if you wish to experiment a bit.

Clearly, the Poisson model is *not* relevant for any of the state or time-series data files on the DATABASE-1 diskette. In the Poisson model, X must be a small nonnegative integer (X = 0, 1, 2, . . .). For example, if X is the number of goals scored by a certain hockey team in the last ten games, your data file would be expected to contain ten numbers such as 0 3 1 1 2 0 0 2 0 1 (except that the numbers would be arranged in a column).

If GOODFT detects large X values, it will recommend using the normal model instead. If GOODFT detects negative or noninteger X values, the program assumes that you have made a mistake and will terminate. This error protection is automatic, though it may not be foolproof. Try to check your data for "reasonableness" before you run GOODFT.

A Poisson model is completely described by one parameter, the mean. From an ungrouped sample in a data file, GOODFT will estimate the sample mean and will tabulate the observed frequency of each X value. If the mean is large, the sample histogram may resemble a bell-shaped curve. If the mean is small, the sample will be strongly skewed to the right.

You may enter the observed frequencies of the X values directly, without a data file. This option is used a lot, since Poisson data often are already grouped into classes (as in some of the exercises at the end of this chapter).

A Poisson test always has an open-ended class on the high end, since technically X has no upper limit. Classes will be combined (collapsed) from the top until expected frequencies become large enough to carry out the calculation. A similar collapsing will be carried out, if necessary, on the low end. If the program cannot collapse the classes sufficiently, you will get an error message, and the program will terminate. If expected frequencies are below 5 in any class, you will receive a warning of possible inflation of the chi-square test statistic.

Combining classes may mean using fewer classes than you hoped. However, a more detailed breakdown simply is not justified unless the sample size is large.

Case Study 11.1: Kentucky Derby Winning Times

Winning Kentucky Derby times for past years are shown in Table 11.1. We use the EXPLORE file editor to create a data file called DERBY. This file contains thirty-four cases, which should be a large enough sample for a normality test.

To see whether these times are normally distributed, we will test the following hypotheses using GOODFT:

H_0: Winning Derby times are from a normal population.

H_1: Winning Derby times are not from a normal population.

From the GOODFT printout (shown on pages 116–117) we see that the summary statistics for the sample suggest little dispersion: The standard de-

Table 11.1   Kentucky Derby Winners 1950–1983

Year	Winning Horse	Time (in seconds)	Year	Winning Horse	Time (in seconds)
1950	Middleground	121.6	1967	Proud Clarion	120.6
1951	Count Turf	122.6	1968	Dancer's Image	122.2
1952	Hill Gail	121.6	1969	Majestic Prince	121.8
1953	Dark Star	122.0	1970	Dust Commander	123.4
1954	Determine	123.0	1971	Canonero II	123.2
1955	Swaps	121.8	1972	Riva Ridge	121.8
1956	Needles	123.4	1973	Secretariat	119.4
1957	Iron Liege	122.2	1974	Cannonade	124.0
1958	Tiny Tam	125.0	1975	Foolish Pleasure	122.0
1959	Tomy Lee	122.2	1976	Bold Forbes	121.6
1960	Venetian Way	122.4	1977	Seattle Slew	122.2
1961	Carry Back	124.0	1978	Affirmed	121.2
1962	Decidedly	120.4	1979	Spectacular Bid	122.4
1963	Chateaugay	121.8	1980	Genuine Risk	122.0
1964	Northern Dancer	120.0	1981	Pleasant Colony	122.0
1965	Lucky Debonair	121.2	1982	Gato del Sol	122.4
1966	Kauai King	122.0	1983	Sunny's Halo	122.2

Source: *Reader's Digest 1982 Almanac* (Pleasantville, N.Y.: Reader's Digest Association, 1982), pp. 832–833; *Facts on File, 1982* (New York: Facts on File, Inc., 1982); *Facts on File, 1983* (New York: Facts on File, Inc., 1983).

viation is small (1.1 seconds), as is the range (5.6 seconds). The coefficient of skewness (.123) and coefficient of kurtosis (3.847) tell us that the sample is not exactly bell-shaped. Although the sample is fairly symmetric (skewness coefficient near zero), it appears that the sample has a sharper peak than a normal distribution (kurtosis statistic above 3). However, Appendix G indicates that the kurtosis coefficient is not large enough to convince us that the population is really leptokurtic.

Despite the suggestion of leptokurtosis, there is little tendency toward a high peak on the standardized histogram. The horses' winning times are distributed fairly symmetrically about the mean: nineteen horses below the mean and fifteen above the mean. Frequencies in corresponding classes above and below the mean are approximately equal, despite a little sample variation.

The histogram shows one horse more than two standard deviations below the mean (Secretariat, in 1973, at 119.4 seconds) and one horse more than two standard deviations above the mean (Tiny Tam, in 1958, at 125.0 seconds). Neither horse is a true outlier, since both are within three standard deviations of the mean. The indications seem favorable for the hypothesis of normality, so far.

To avoid small expected frequencies, the computer has combined the end classes to yield only four classes ($k = 4$). Since the mean and standard deviation were estimated ($m = 2$), degrees of freedom are

$$\text{d.f.} = k - m - 1 = 4 - 2 - 1 = 1.$$

From Appendix D the critical value of chi-square for the .05 level of significance is 3.841, so the decision rule is as shown in Figure 11.3.

Since the test statistic (1.604) does not exceed the critical value (3.841), the hypothesis of normality cannot be rejected. Given the rest of the supporting evidence, we conclude that our sample of winning Kentucky Derby times may be from a normal population.

We have not actually proved normality; rather, we have failed to disprove it. This is a subtle but possibly important distinction to bear in mind.

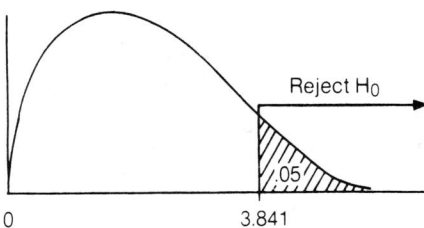

Figure 11.3 Chi-Square Test for Normality

GOODFT

This program employs the chi-square statistic to test a
set of observed data for goodness-of-fit to various
theoretical probability distributions. If you have
ungrouped data, they should be contained in a saved
file consisting of a single column of N observations
where N is up to 1000. These abbreviations are used:

           g-o-f = goodness-of-fit
           f(o) = observed frequency
           f(e) = expected frequency

Four g-o-f tests are available:

               1. Normal Model
               2. Uniform Model
               3. Poisson Model
               4. Do-It-Yourself Test

Which g-o-f test (1,2,3, or 4)? 1
File name? derby

```
===================================================
                 SUMMARY STATISTICS
---------------------------------------------------
   File = DERBY
  # Obs =   34
    Low =  119.4
   High =  125
   Mean =  122.1
 Median =  122
St.Dev.=  1.117627    (using sample definition)
St.Dev.=  1.101069    (using universe definition)
Skewness =   0.123   (Pearson skewness coefficient)
Kurtosis =   3.847   (Pearson kurtosis coefficient)
---------------------------------------------------
```

```
  f(o)     f(e)     Z Scale and Histogram
 ------   ------    +-------------------------
    0      0.0      |
                  -3+
    1      0.7      | *
                  -2+
    3      4.6      | ***
                  -1+
   15     11.6      | ***************
                   0+
   10     11.6      | *********
                   1+
    4      4.6      | ****
                   2+
    1      0.7      | *
                   3+
    0      0.0      |
```

```
=============================================================
Standardized                                              2
Z-value        f(o)    f(e)    f(o)-f(e)    [f(o)-f(e)] /f(e)
-------------------------------------------------------------
Under -1.0      4      5.4      -1.4         0.360
-1.0 to 0.0    15     11.6       3.4         0.993
0.0 to 1.0     10     11.6      -1.6         0.222
1.0 and Over    5      5.4      -0.4         0.029
-------------------------------------------------------------
Total:         34     34.0       0.0         1.604

Chi-Square =    1.604  with d.f.=  1
Chi-Square =    0.978  using Yates' correction

Note: Details may not add to totals, due to rounding.
      The first 3 and last 3 classes were combined
      to enlarge f(e) for file DERBY.

Where now:  E=EXPLORE menu   R=Run GOODFT again   X=exit ? e
```

■ Case Study 11.2: Presidents' Ages at Inauguration

The ages at inauguration of the first forty U.S. presidents are shown in Table 11.2. After using the EXPLORE file editor to create a data file named PRES-AGE, we can run the program GOODFT.

The sample should be large enough for a valid test of these hypotheses:

H_0: U.S. Presidents' ages follow a uniform distribution.

H_1: U.S. Presidents' ages do not follow a uniform distribution.

From Table 11.2 and from the printout's summary statistics we see that the range is from age 42 (T. Roosevelt) to age 70 (Reagan). The skewness coefficient (.287) is near zero, indicating near-symmetry. However, the kurtosis statistic (3.017) is strongly suggestive of a normal population (a uniform population's kurtosis coefficient would be near 1.8, while a normal population's kurtosis coefficient would be near 3.0). There are no extreme outliers, since even the extreme points (T. Roosevelt and Reagan) are within three standard deviations of the mean.

The computer seems to have chosen appropriate "aesthetically pleasing" class limits, as it is supposed to. It is quite clear from the table of frequencies that the presidents' ages at inauguration are not uniformly distributed. If we were to sketch a histogram using the classes chosen by the computer, it would look bell-shaped—but certainly not rectangular.

Since preliminary evidence casts doubt on the hypothesis that the sample is from a uniform population, we anticipate that the test statistic will be large and will lead us to reject H_0. Since there are four classes (k = 4) and no parameters are estimated (m = 0), our degrees of freedom will be

$$d.f. = k - m - 1 = 4 - 0 - 1 = 3.$$

From Appendix D, using the .05 level of significance and d.f. = 3, we find a critical value of chi-square and set up our decision rule as shown in Figure 11.4.

Table 11.2 Presidents' Ages at Inauguration

President	Age	President	Age
1. Washington	57	21. Arthur	50
2. J. Adams	61	22. Cleveland	47
3. Jefferson	57	23. B. Harrison	55
4. Madison	57	24. Cleveland	55
5. Monroe	58	25. McKinley	54
6. J. Q. Adams	57	26. T. Roosevelt	42
7. Jackson	61	27. Taft	51
8. Van Buren	54	28. Wilson	56
9. W. H. Harrison	68	29. Harding	55
10. Tyler	51	30. Coolidge	51
11. Polk	49	31. Hoover	54
12. Taylor	64	32. F. Roosevelt	51
13. Fillmore	50	33. Truman	60
14. Pierce	48	34. Eisenhower	62
15. Buchanan	65	35. Kennedy	43
16. Lincoln	52	36. L. Johnson	55
17. A. Johnson	56	37. Nixon	56
18. Grant	46	38. Ford	62
19. Hayes	54	39. Carter	52
20. Garfield	49	40. Reagan	70

Source: *Hammond Almanac, 1982* (Maplewood, N.J.: Hammond Almanac, Inc., 1982), pp. 140–141.

Since the test statistic (29.000) greatly exceeds the critical value (7.815), it is easy to reject the hypothesis that presidents' ages at inauguration are uniformly distributed. It would be a good idea to rerun GOODFT using the normal model, which might give a better fit.

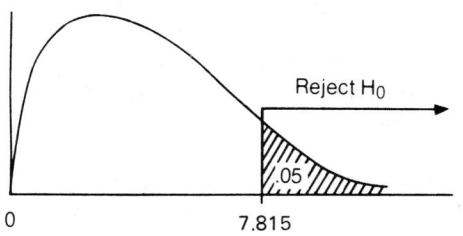

Figure 11.4 Chi-Square Test for Uniformity

GOODFT

This program employs the chi-square statistic to test a
set of observed data for goodness-of-fit to various
theoretical probability distributions. If you have
ungrouped data, they should be contained in a saved
file consisting of a single column of N observations
where N is up to 1000. These abbreviations are used:

       g-o-f = goodness-of-fit
       f(o) = observed frequency
       f(e) = expected frequency

Four g-o-f tests are available:

       1. Normal Model
       2. Uniform Model
       3. Poisson Model
       4. Do-It-Yourself Test

Which g-o-f test (1,2,3, or 4)? 2
Are your observations saved in a data file (y/n)? y
File name? presage

```
====================================================
               SUMMARY STATISTICS
----------------------------------------------------
   File = PRESAGE
 # Obs =   40
   Low =   42
  High =   70
  Mean =   54.875
Median =   55
St.Dev.=  6.227267   (using sample definition)
St.Dev.=  6.148933   (using universe definition)
Skewness =   0.287   (Pearson skewness coefficient)
Kurtosis =   3.017   (Pearson kurtosis coefficient)
----------------------------------------------------
```

              (Using 4 Classes)

```
==================================================
From      To (Not Incl)       f(o)       f(e)
--------------------------------------------------
  40          50               7        10.000
  50          60              24        10.000
  60          70               8        10.000
  70          80               1        10.000
--------------------------------------------------
Total:                        40        40.000
```

CALCULATION OF TEST STATISTIC FOR FILE PRESAGE

$$f(o)-f(e) \qquad [f(o)-f(e)]^2 /f(e)$$

```
  ---------      ---------------
    -3.000            0.900
    14.000           19.600
    -2.000            0.400
    -9.000            8.100
  ---------      ---------------
     0.000           29.000
```

Chi-Square =   29.000
      d.f. =   3

Where now: E=EXPLORE menu   R=Run GOODFT again   X=exit ? e

■ Case Study 11.3: Highway Fatalities by Day of Week

On the evening news we often hear reports of the "high weekend highway death toll." Do weekends really have more traffic fatalities? Consider the statistics shown in Table 11.3.

We can see that weekdays are apparently safer, based on this sample of one month's data. But is the distribution *significantly* different from a uniform model? The hypotheses are

H_0: Traffic deaths are uniformly distributed throughout the days of the week.

H_1: Traffic deaths are not uniformly distributed throughout the days of the week.

We do not use a data file, since the frequencies are already calculated. Instead, we just enter the seven observed frequencies while GOODFT is running. The program prompts us with a series of question marks (see the sample printout on pages 121–122). If we make an error, we can just let the program run, then rerun it with the correct data. We do *not* enter the total from the bottom of the column—the computer will add the numbers.

If the traffic fatalities had been uniformly distributed, 118 fatalities per day would be expected. While some days are close to 118 fatalities, Saturday (with 182 fatalities) is far above the expected uniform frequency. The first four days of the week are consistently below 118, and the weekend days are consistently too high. Because of the relatively large sample size (n = 826 fatalities), the observed departure from a uniform distribution is likely to be significant, and we anticipate a large chi-square test statistic.

There are seven classes (k = 7), and no parameters were estimated (m = 0), so

$$\text{d.f.} = k - m - 1 = 7 - 0 - 1 = 6.$$

Table 11.3   U.S. Traffic Deaths, January 1980

Day of Week	Number of Deaths
Monday	91
Tuesday	91
Wednesday	99
Thursday	99
Friday	132
Saturday	182
Sunday	132
Total:	826

Source: *Accident Facts, 1981* (Chicago: National Safety Council, 1981), p. 51.

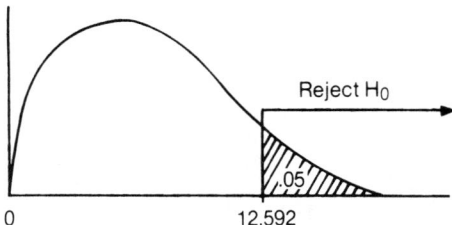

Figure 11.5   Chi-Square Test for Uniformity

The decision rule is shown in Figure 11.5, using Appendix D to obtain the critical value of chi-square at the .05 level of significance.

Since the sample test statistic (56.508) greatly exceeds the critical value (12.592), the decision is not even close. Clearly, the hypothesis of uniformly distributed traffic deaths must be rejected. Perhaps you can list the various reasons *why* there are more traffic fatalities on the weekend than on weekdays.

```
                          GOODFT

       This program employs the chi-square statistic to test a
       set of observed data for goodness-of-fit to various
       theoretical probability distributions.  If you have
       ungrouped data, they should be contained in a saved
       file consisting of a single column of N observations
       where N is up to 1000.   These abbreviations are used:

                  g-o-f = goodness-of-fit
                   f(o) = observed frequency
                   f(e) = expected frequency

       Four g-o-f tests are available:

                     1. Normal Model
                     2. Uniform Model
                     3. Poisson Model
                     4. Do-It-Yourself Test

       Which g-o-f test (1,2,3, or 4)? 2
       Are your observations saved in a data file (y/n)? n


       FREQUENCIES FOR UNIFORM TEST

       How many classes for the uniform test? 7

       When requested, enter your observed
       frequency f(o) for each class.
       Class           f(o)
       -----           ----
         1             ? 91
         2             ? 91
         3             ? 99
         4             ? 99
         5             ? 132
         6             ? 182
         7             ? 132
```

```
              (Using   7  Classes)

     f(o)           f(e)
     ----           ----
      91          118.000
      91          118.000
      99          118.000
      99          118.000
     132          118.000
     182          118.000
     132          118.000
     ----           ----
     826          826.000
```

```
     CALCULATION OF TEST STATISTIC

                              2
     f(o)-f(e)     [f(o)-f(e)] /f(e)
     ---------     ----------------
      -27.000            6.178
      -27.000            6.178
      -19.000            3.059
      -19.000            3.059
       14.000            1.661
       64.000           34.712
       14.000            1.661
     ---------     ----------------
        0.000           56.508

     Chi-Square =    56.508
           d.f. =     6

     Where now:  E=EXPLORE menu   R=Run GOODFT again   X=exit ? e
```

■ Case Study 11.4: Supreme Court Vacancies

The number of Supreme Court seats that become vacant in a given year might be hypothesized to be a Poisson variable, since "rare" events occurring independently over time are often well approximated by the Poisson model. We formulate these hypotheses:

H_0: Annual U.S. Supreme Court vacancies follow a Poisson model.

H_1: Annual U.S. Supreme Court vacancies do not follow a Poisson model.

The number of vacancies in each year from 1900 to 1982 are shown in Table 11.4. This sample of eighty-three observations should be large enough to obtain a valid hypothesis test.

Before using the computer we note that the range of vacancies is rather small. Only twice have there been as many as three vacancies in one year (1910 and 1922), and most often there are none.

Table 11.4   Number of U.S. Supreme Court Vacancies, 1900–1982

Year	Vacancies	Year	Vacancies	Year	Vacancies
1900	0	1928	0	1956	1
1901	0	1929	0	1957	1
1902	1	1930	2	1958	1
1903	1	1931	0	1959	0
1904	0	1932	1	1960	0
1905	0	1933	0	1961	0
1906	1	1934	0	1962	2
1907	0	1935	0	1963	0
1908	0	1936	0	1964	0
1909	1	1937	1	1965	1
1910	3	1938	2	1966	0
1911	1	1939	2	1967	1
1912	0	1940	0	1968	0
1913	0	1941	2	1969	2
1914	1	1942	1	1970	0
1915	0	1943	0	1971	2
1916	1	1944	0	1972	0
1917	0	1945	1	1973	0
1918	0	1946	1	1974	0
1919	0	1947	0	1975	1
1920	0	1948	0	1976	0
1921	1	1949	2	1977	0
1922	3	1950	0	1978	0
1923	0	1951	0	1979	0
1924	0	1952	0	1980	0
1925	1	1953	1	1981	1
1926	0	1954	1	1982	0
1927	0	1955	0		

Source: *Hammond Almanac, 1982* (Maplewood, N.J.: Hammond Almanac, Inc., 1982), pp. 180–182.

We could use the EXPLORE file editor to create a new data file containing these eighty-three observations. However, the data are simple integers. Rather than typing eighty-three data points by hand, it is easier to tabulate the data first. Table 11.5 shows the summary of the raw data, which we will use when we run GOODFT.

Table 11.5   U.S. Supreme Court Vacancies, 1900–1982

Number of Vacancies	Number of Years
0	50
1	23
2	8
3	2
Total:	83

Using GOODFT, we generate the printout (shown on pages 125–126). We do not use a data file because the frequencies are already calculated. Instead, we just enter the observed frequencies from Table 11.5 while GOODFT is running. We have to tell GOODFT that the largest data value to expect is 3 so it will know when to stop prompting us for input. The program prompts us with a series of question marks. If we make an error, we can just let the program run, then rerun it with the correct data. We do *not* enter the total from the bottom of the column—the computer will add the numbers.

To enlarge f(e), GOODFT combines the top two classes to form a class called "2 or more." Based on an eyeball comparison of f(o) and f(e), we can see that the Poisson model gives a very good fit to the court vacancies. We expect that the test statistic will be very near zero and that we will not be able to reject H_0 in the formal hypothesis tests.

There are three classes (k = 3) and one parameter is estimated (m = 1), so degrees of freedom will be

$$d.f. = k - m - 1 = 3 - 1 - 1 = 1.$$

Therefore from Appendix D the critical value of chi-square at the .05 level of significance with d.f. = 1 gives the decision rule shown in Figure 11.6.

Since the continuity-corrected chi-square test statistic (.404) is far below the critical value (3.841), we cannot reject H_0. We conclude that annual U.S. Supreme Court vacancies do follow a Poisson distribution.

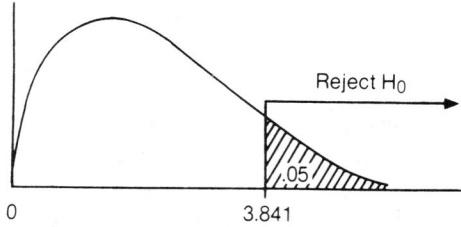

Figure 11.6   Chi-Square Test for Poisson Model

GOODFT

This program employs the chi-square statistic to test a set of observed data for goodness-of-fit to various theoretical probability distributions. If you have ungrouped data, they should be contained in a saved fie consisting of a single column of N observations where N is up to 1000. These abbreviations are used:

```
g-o-f = goodness-of-fit
f(o) = observed frequency
f(e) = expected frequency
```

Four g-o-f tests are available:

```
1. Normal Model
2. Uniform Model
3. Poisson Model
4. Do-It-Yourself Test
```

Which g-o-f test (1,2,3, or 4)? 3
Are your observations saved in a data file (y/n)? n

ENTRY OF OBSERVED X FREQUENCIES

You'll enter the observed frequency of X in this format:

```
X    f(o)
---  ----
0    xx
1    xx
2    xx
:    :
:    :
m    xx     where m is the highest X value in your data.
```

Enter the value of m? 3

```
X    f(o)
---  ----
0    ? 50
1    ? 23
2    ? 8
3    ? 2
```

... POISSON GOODNESS OF FIT TEST ...

Mean =   0.54 events per unit of time/space
          (rounded to nearest .01)

X	P(X)	f(o)	f(e)
0	0.5815	50	48.3
1	0.3153	23	26.2
2	0.0855	8	7.1
3 and over	0.0178	2	1.5
Totals:	1.0000	83	83.0

```
DETAILS OF POISSON TEST STATISTIC CALCULATION

                                    2
      X              [f(o) - f(e)] /f(e)
    -----           -----------------
      0                    0.062
      1                    0.33
    2 and more             0.239
    -----           -----------------
    Total:                 0.684

    Chi-Square =    0.684 with d.f. =  1
    Chi-Square =    0.404 using Yates correction

    Where now:  E=EXPLORE menu   R=Run GOODFT again   X=exit ? e
```

■ Case Study 11.5: Checking RAND's Poisson Data

Using RAND (see RAND's Case Study 19.4), we generate one hundred random numbers using the Poisson model with a mean of 2.5. These numbers are saved in a data file named POIS33. According to statistical theory (see RAND and POIS for details), the true parameters of RAND's sample *should* have been

$$\text{Mean} = \mu = 2.5$$
$$\text{S.D.} = \sqrt{\mu} = \sqrt{2.5} = 1.581$$
$$\text{Skewness} = 1/\sqrt{\mu} = 1/\sqrt{2.5} = .632$$
$$\text{Kurtosis} = 3 + 1/\mu = 3 + 1/2.5 = 3.400.$$

To see what actually happened, we use GOODFT to analyze the data file POIS33. As might be anticipated, RAND's sample of one hundred Poisson random numbers does not exactly yield the theoretical statistics. However, the printout shows that the sample statistics are pretty close to the intended values:

	Sample	Desired
Mean	2.59	2.50
S.D.	1.609	1.581
Skewness	.729	.632
Kurtosis	3.051	3.400

On the basis of these statistics it looks as if RAND has generated a sample possessing the desired Poisson characteristics.

However, we must also look at the formal goodness-of-fit test. The smallest data value is 0, and the largest is 7 (see printout). The observed and expected frequencies seem fairly close, except that there are not quite enough 4s. Degrees of freedom for the test will be

$$\text{d.f.} = k - 2 = 7 - 2 = 5.$$

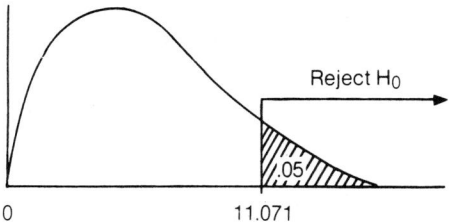

Figure 11.7   Chi-Square Test for Poisson Model

The last two classes were collapsed by GOODFT to increase the expected frequency in the last class. At the .05 level of significance the critical value of chi-square is 11.071, from Appendix D. From the printout the chi-square test statistic is 4.800, so we cannot reject the hypothesis that the data file POIS33 contains Poisson random numbers. RAND seems to be working as intended. The decision rule is illustrated in Figure 11.7.

```
                            GOODFT

        This program employs the chi-square statistic to test a
        set of observed data for goodness-of-fit to various
        theoretical probability distributions.  If you have
        ungrouped data, they should be contained in a saved
        file consisting of a single column of N observations
        where N is up to 1000.  These abbreviations are used:

                 g-o-f = goodness-of-fit
                 f(o) = observed frequency
                 f(e) = expected frequency

        Four g-o-f tests are available:

                 1. Normal Model
                 2. Uniform Model
                 3. Poisson Model
                 4. Do-It-Yourself Test

        Which g-o-f test (1,2,3, or 4)? 3
        Are your observations saved in a data file (y/n)? y
        File name? pois33


        =====================================================
                         SUMMARY STATISTICS
        -----------------------------------------------------
           File = POIS33
          # Obs =  100
            Low =  0
           High =  7
           Mean =  2.59
         Median =  2
        St.Dev.=  1.608657   (using sample definition)
        St.Dev.=  1.600593   (using universe definition)
        Skewness =   0.729   (Pearson skewness coefficient)
        Kurtosis =   3.051   (Pearson kurtosis coefficient)
        -----------------------------------------------------
```

```
Mean =    2.59 events per unit of time/space
              (rounded to nearest .01)
```

X	P(X)	f(o)	f(e)
0	0.0750	5	7.5
1	0.1943	22	19.4
2	0.2516	28	25.2
3	0.2172	22	21.7
4	0.1407	8	14.1
5	0.0729	9	7.3
6	0.0315	4	3.1
7 and over	0.0169	2	1.7
Totals:	1.0000	100	100.0

```
DETAILS OF POISSON TEST STATISTIC
CALCULATION FOR FILE POIS33
```

X	$[f(o) - f(e)]^2/f(e)$
0	0.834
1	0.340
2	0.320
3	0.004
4	2.616
5	0.403
6 and more	0.283
Total:	4.800

```
Chi-Square =   4.800 with d.f. =   5

Where now:  E=EXPLORE menu   R=Run GOODFT again   X=exit ? e
```

■ Case Study 11.6: Checking DICE

Referring back to Case Study 10.1 (the DICE program), we can use GOODFT to carry out a goodness-of-fit test on the frequencies that we encountered in DICE. It will be recalled that some of the frequencies seemed different from what would be expected. If these differences are not within chance, we might suspect that something was wrong with the DICE program. For seventy-two rolls of two dice (see DICE printout) we received the results shown in Table 11.6.

The hypotheses of interest are

H_0:   Sum of two dice follows the correct theoretical model (see DICE case study).

H_1:   Sum of two dice differs significantly from the correct theoretical model.

We use the Do-It-Yourself option of GOODFT to enter the actual and expected frequencies. We do not use any files. Instead, the frequencies are typed

Table 11.6   Results for 72 Rolls of Two Dice

Dots Showing	Actual Frequency	Expected Frequency
2	2	2
3	1	4
4	3	6
5	4	8
6	13	10
7	18	12
8	14	10
9	7	8
10	7	6
11	3	4
12	0	2
Total:	72	72

in as GOODFT requests. If we make a typing error, we can just rerun the program.

Some of the expected frequencies are rather small (see sample printout). We should remember this as a possible limitation, although hopefully not a severe problem. The computer cannot determine the degrees of freedom, because it does not know what model we are using. However, the calculation of the degrees of freedom is quite simple. There are eleven classes (k = 11) and no parameters estimated (m = 0), so

$$\text{d.f.} = k - m - 1 = 11 - 0 - 1 = 10.$$

We decline the use of Yates's continuity correction (it would be used only if d.f. = 1). At the .05 level of significance, Appendix D shows that the critical chi-square value for d.f. = 10 is 18.307. Figure 11.8 shows the decision rule.

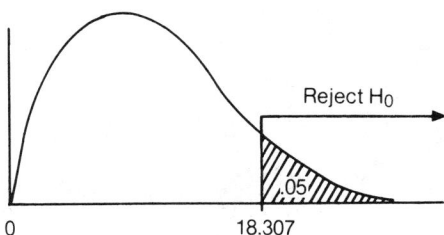

Figure 11.8   Chi-Square Test for Dice Rolls

From the GOODFT printout the test statistic (13.792) is not large enough to reject H_0, since it does not exceed the critical value (18.307). However, further scrutiny of the DICE random number generator is desirable, since the test statistic is nearly large enough to have been significant at the .10 level. A larger sample size might be tried, say by rolling the dice 144 times instead of 72 times. If the pattern persists, the program DICE should be dissected to find the underlying cause.

```
                        GOODFT

        This program employs the chi-square statistic to test a
        set of observed data for goodness-of-fit to various
        theoretical probability distributions.  If you have
        ungrouped data, they should be contained in a saved
        file consisting of a single column of N observations
        where N is up to 1000.  These abbreviations are used:

                    g-o-f = goodness-of-fit
                    f(o) = observed frequency
                    f(e) = expected frequency

        Four g-o-f tests are available:

                    1. Normal Model
                    2. Uniform odel
                    3. Poisson Model
                    4. Do-It-Yourself Test

        Which g-o-f test (1,2,3, or 4)? 4



            ... DO-IT-YOURSELF GOODNESS OF FIT TEST ...

            The computer does not specify the probability
        model .  You must provide f(o) and f(e) for each class.
        In other words, prior grouped data are assumed to exist.

        How many classes? 11
        Enter f(o), f(e) for each class,
        (separated by comma e.g. 9,6):

        Class           f(o),f(e)
        -----           ---------
          1             ? 2,2
          2             ? 1,4
          3             ? 3,6
          4             ? 4,8
          5             ? 13,10
          6             ? 18,12
          7             ? 14,10
          8             ? 7,8
          9             ? 7,6
         10             ? 3,4
         11             ? 0,2
```

```
Yates' correction should be used when d.f. = 1 where
d.f. = number of classes - number of parameters. Do you
want me to use Yates' continuity correction (y/n)? n


    GOODNESS OF FIT TEST FOR GROUPED DATA:
                                                2
    f(o)      f(e)     f(o)-f(e)      [f(o)-f(e)] /f(e)
    ----      ----     ---------      ----------------
      2        2.0          0             0.000
      1        4.0         -3             2.250
      3        6.0         -3             1.500
      4        8.0         -4             2.000
     13       10.0          3             0.900
     18       12.0          6             3.000
     14       10.0          4             1.600
      7        8.0         -1             0.125
      7        6.0          1             0.167
      3        4.0         -1             0.250
      0        2.0         -2             2.000

     72        72           0            13.792

Chi-Square =  13.792
Warning--Some of the f(e) are less than 5.

Where now:  E=EXPLORE menu   R=Run GOODFT again   X=exit ? e
```

■ Exercises

11.1   Use GOODFT's Option 1 to test whether the years served by the popes
       since 1500 A.D. (see Table 11.7) follow a normal distribution. Create your
       data file POPES, using the EXPLORE file editor. Logically, would you
       a priori expect the data to be normal? Why or why not? Discuss what the
       eyeball test of the histogram tells you about the shape of the distribution.
       Verify the degrees of freedom, and choose a level of significance. Set up
       a decision rule. Illustrate. Is the decision sensitive to the level of signifi-
       cance chosen? Discuss fully.

11.2   Use GOODFT's Option 2 to test whether the season in which the first
       thirty-six U.S. Presidents died (see Table 11.8) follows a uniform distri-
       bution. Logically speaking, would you expect the deaths to be uniform *a
       priori?* Why or why not? Make a histogram. Discuss what the eyeball test
       of the histogram tells you about the shape of the distribution. Verify the
       degrees of freedom, and choose a level of significance. Set up a decision
       rule. Illustrate. Is the decision sensitive to the level of significance chosen?
       Discuss fully.

11.3   Choose one of the state data files from the DATABASE-1 diskette, and
       repeat Exercise 11.1.

11.4   Choose one of the state data files from the DATABASE-1 diskette, and
       repeat Exercise 11.2.

Table 11.7   Years Served by Roman Catholic Popes Since 1500

Pope	Service	Years	Pope	Service	Years
Pius III	1503	0	Innocent XI	1676–1689	13
Julius II	1503–1513	10	Alexander VIII	1689–1691	2
Leo X	1513–1521	8	Innocent XII	1691–1700	9
Adrian VI	1522–1523	1	Clement XI	1700–1721	21
Clement VII	1523–1534	11	Innocent XIII	1721–1724	3
Paul III	1534–1549	15	Benedict XIII	1724–1730	6
Julius III	1550–1555	5	Clement XII	1730–1740	10
Marcellus II	1555	0	Benedict XIV	1740–1758	18
Paul IV	1555–1559	4	Clement XIII	1758–1769	11
Pius IV	1559–1565	6	Clement XIV	1769–1774	5
St. Pius V	1566–1572	6	Pius VI	1775–1799	24
Gregory XIII	1572–1585	13	Pius VII	1800–1823	23
Sixtus V	1585–1590	5	Leo XII	1823–1829	6
Urban VII	1590	0	Pius VIII	1829–1830	1
Gregory XIV	1590–1591	1	Gregory XVI	1831–1846	15
Innocent IX	1591	0	Pius IX	1846–1878	32
Clement VIII	1592–1605	13	Leo XIII	1878–1903	25
Leo XI	1605	0	St. Pius X	1903–1914	11
Paul V	1605–1621	16	Benedict XV	1914–1922	8
Gregory XV	1621–1623	2	Pius XI	1922–1939	17
Urban VIII	1623–1644	21	Pius XII	1939–1958	19
Innocent X	1644–1655	11	John XXIII	1958–1963	5
Alexander VII	1655–1667	12	Paul VI	1963–1978	15
Clement IX	1667–1669	2	John Paul I	1978	0
Clement X	1670–1676	6	John Paul II	1978–	

Source: *Reader's Digest 1982 Almanac* (Pleasantville, N.Y.: Reader's Digest Association, 1982), p. 709.

11.5   Assume that a supposedly fair coin is flipped one hundred times, coming up "heads" sixty times. Use GOODFT's Option 2 to test whether the coin really is fair. What potential problem exists with degrees of freedom? Explain your test. Could you also use a z test for one proportion to test p = .5? What difference is there? Which test is more powerful?

Table 11.8 Season During Which the First Thirty-Six U.S. Presidents Died

Season	Deaths
January–March	11
April–June	9
July–September	10
October–December	6
Total:	36

Source: *Hammond Almanac, 1982* (Maplewood, N.J.: Hammond Almanac, Inc., 1982), pp. 140–141.

11.6 Assume that a supposedly fair die is rolled sixty times, resulting in the distribution of dots shown in Table 11.9. Use GOODFT's Option 2 to test whether the die is fair. Explain fully, showing all steps in your reasoning.

Table 11.9 Dots Showing on One Die Rolled Sixty Times

Dots	1	2	3	4	5	6	Total
Frequency	8	14	9	12	7	10	60

11.7 During the 1973–1974 hockey season the Boston Bruins played thirty-nine home games and scored 193 points, as shown in Table 11.10. Use GOODFT's Option 3 to test whether the goals scored can be said to follow a Poisson distribution. Logically speaking, would you expect the goals to be Poisson *a priori?* Why or why not? Make a histogram. Discuss what the eyeball test of the histogram tells you about the shape of the distribution. Verify the degrees of freedom, and choose a level of significance. Set up a decision rule. Illustrate. Is the decision sensitive to the level of significance chosen? Discuss fully.

Optional Exercises for Ambitious Learners

11.8 Run RAND to generate one hundred normally distributed random numbers with a mean of 0 and a standard deviation of 1. Save the random numbers in a file called NORM1. These random numbers should be symmetric (skewness coefficient of 0) and mesokurtic (kurtosis coefficient of 3). Moreover, they should give a chi-square test statistic near 0 when you try GOODFT's Option 1. Find out what really happens. Is RAND performing properly? Use Appendixes F and G to interpret the skewness and kurtosis coefficients. Discuss fully, including the mean and standard deviation of the sample.

Table 11.10   Goals Scored by Boston Bruins, 1973–1974

Number of Goals	Number of Games
0	0
1	1
2	2
3	5
4	9
5	10
6	5
7	2
8	3
9	1
10	1
Total:	39

Source: Gary M. Mullett, "Simeon Poisson and the National Hockey League," *The American Statistician*, 31 (February, 1977): 9.

11.9   Run RAND to generate one hundred uniformly distributed random numbers between 0 and 1999. Save the random numbers in a file called UNIF1. These random numbers should be symmetric (skewness coefficient of 0) and platykurtic (kurtosis coefficient of 1.8). Moreover, they should give a chi-square test statistic near 0 when you try GOODFT's Option 2. Find out what really happens. Is RAND performing properly? Use Appendixes F and G to interpret the skewness and kurtosis coefficients. Discuss fully, including the mean and standard deviation of the sample.

11.10   Run RAND to generate one hundred Poisson random numbers with a mean of 5. Save the random numbers in a file called POIS1. These random numbers should be skewed to the right (skewness coefficient significantly greater than 0) and leptokurtic (kurtosis coefficient greater than 3). Moreover, they should give a chi-square test statistic near 0 when you try GOODFT's Option 3. Find out what really happens. Is RAND performing properly? Use Appendixes F and G to interpret the skewness and kurtosis coefficients. Discuss fully, including the mean and standard deviation of the sample.

11.11   The St. Loius Cardinals won the 1982 World Series against the Milwaukee Brewers. Table 11.11 shows the inning-by-inning breakdown of runs by each team. It could be argued that spectator excitement is related to the number of runs scored per inning. Complete the blank table to test whether X = runs per inning is a Poisson variable. Do this two ways. First, use n = 58 innings (defining X as "total runs per inning by both teams").

Then use n = 121 (defining X as "runs per inning with each team at bat treated as a separate event"). Use Option 3 of GOODFT to perform the calculations on the data that you have tabulated. That is, do two separate chi-square tests for Poisson distribution, using your two separate tabulations.

Table 11.11   1982 World Series Box Scores
Key: Mil = Milwaukee
StL = St. Louis

| | | \multicolumn{9}{c}{Inning} | Total |
		1	2	3	4	5	6	7	8	9	Runs:
Game 1:	Mil	2	0	0	1	1	2	0	0	4	10
	StL	0	0	0	0	0	0	0	0	0	0
Game 2:	Mil	0	1	2	0	1	0	0	0	0	4
	StL	0	0	2	0	0	2	0	1	x	5
Game 3:	StL	0	0	0	0	3	0	2	0	1	6
	Mil	0	0	0	0	0	0	0	2	0	2
Game 4:	StL	1	3	0	0	0	1	0	0	0	5
	Mil	0	0	0	0	1	0	6	0	x	7
Game 5:	StL	0	0	1	0	0	0	1	0	2	4
	Mil	1	0	1	0	1	0	1	2	x	6
Game 6:	Mil	0	0	0	0	0	0	0	0	1	1
	StL	0	2	0	3	2	6	0	0	x	13
Game 7:	Mil	0	0	0	0	1	2	0	0	0	3
	StL	0	0	0	1	0	3	0	2	x	6

Note: x shows unnecessary last half of inning
Source: *Detroit Free Press,* October 22, 1982, p. 4D.

Runs Scored	0	1	2	3	4	5	6	Total
Frequency (Number of Innings)								58
Frequency (Number of Innings)								121

11.12 In the 1982 World Series a total of 72 runs were scored. It is reasonable to ask whether these runs were uniformly distributed among the nine innings or whether they were concentrated in certain innings. Discuss *a priori* which might be the case and why. Then complete the tabulation below by counting the frequency of runs in each of the nine innings, with all games combined. Then use Option 2 of GOODFT to test the hypothesis that runs are uniformly distributed. Explain your reasoning fully.

Inning	1	2	3	4	5	6	7	8	9	Total
Runs Scored										72

12 GROUP/Grouped Data Analysis

Introduction

Sometimes we must work with data that have been categorized into groups. When any data set is collapsed into classes, we lose information. But we also gain in clarity of presentation because grouped data are often easier to assimilate than raw data.

From grouped data we can usually estimate the mean and standard deviation, as well as the quartile points and certain other statistics. However, our grouped estimates will differ from the values that we obtain from raw data. The accuracy of the grouped estimates will depend on the number of classes, distribution of data within classes, class frequencies, and class sizes.

By hand, the calculations from grouped data can be tedious. The program GROUP not only handles the calculations, but also prints out a worksheet showing the intermediate steps. This helps us to see what is going on and simplifies any further calculations that might be intended.

■ Case Study 12.1: Mutual Fund Performance

Consider Table 12.1, which shows twelve-month rates of return for the period ended April 1, 1986, for fifty growth-oriented mutual funds chosen at random from a population of 183 growth-oriented mutual funds. The methodology was to select every fourth growth-oriented company on each page of a thirteen-page list, starting with the first company on each page.

The input of this data into GROUP is quite simple. Since we have equal classes, we just need to tell GROUP how many classes to use, the interval width, and the starting point.

When calculating a mean or standard deviation from grouped data, we treat all observations within a class as if they were located at the midpoint. For example, in the third class, "20 but less than 30," we pretend that all 23 companies had a 25 percent rate of return. In fact, the observations will generally be scattered throughout each interval. But it makes sense to assume that the data are distributed uniformly within an interval so that *on the average*

Table 12.1   Performance of Fifty Growth-Oriented Mutual Funds

12-Month Return (percent)	Number of Funds
0 but less than 10	1
10 but less than 20	5
20 but less than 30	23
30 but less than 40	13
40 but less than 50	6
50 but less than 60	2
Total:	50

Caution: These data are intended solely for classroom purposes and should not be viewed as a guide to investments.
Source: *Money,* Vol. 15, No. 5, May 1986, pp. 208–234.

they are located at the class midpoint. We could think of assumptions that might work better, particularly in the tails of the histogram, but they would be rather complex.

GROUP calculates the estimated mean by multiplying the midpoint of each class by its class frequency, taking the sum over all classes, and dividing by sample size:

$$\text{Estimated mean} = \Sigma fm/n = 1490/50 = 29.8.$$

GROUP then estimates the standard deviation by subtracting the estimated mean from each class midpoint, squaring the difference, multiplying by the

```
                        GROUP

This program calculates various sample statistics such as the
mean, median, standard deviation, and coefficient of variation
from grouped data.  Closed-end classes (up to 15) are assumed.
You will be asked to input the number of classes, interval
size, class lower limit(s), and frequencies for each class (up
to N=9999). Detailed calculations will be shown on the screen.

How many classes (up to 15)? 6
Equal interval size (y/n)? y
Size of each interval? 10
Lower limit of first class? 0

INPUT FREQUENCIES FOR EACH CLASS INTERVAL

Class               Frequency
---------           ----------
    0 <  10      ? 1
   10 <  20      ? 5
   20 <  30      ? 23
   30 <  40      ? 13
   40 <  50      ? 6
   50 <  60      ? 2
```

SUMMARY OF CALCULATIONS USING GROUPED DATA

Class	f	m	f*m	f*(m-x̄)²	Cum f
0 < 10	1	5	5	615.0400	1
10 < 20	5	15	75	1095.2000	6
20 < 30	23	25	575	529.9198	29
30 < 40	13	35	455	351.5201	42
40 < 50	6	45	270	1386.2401	48
50 < 60	2	55	110	1270.0801	50
Total:	50		1490	5248.0000	

SUMMARY STATISTICS FOR GROUPED DATA

Mean = 29.8
St. Dev. = 10.34901 (using sample formula)
St. Dev. = 10.245 (using population formula)
Coef. of Variation = 34.7 (using sample formula)
Coef. of Variation = 34.4 (using population formula)
1st Quartile = 22.82609
2nd Quartile = 28.26087
3rd Quartile = 36.53846

Where now: E=EXPLORE menu   R=Run GROUP again   X=exit ? e

class frequency, taking the sum over all classes to obtain the sum of squared deviations about the mean, dividing by n − 1 (or n for a population), and taking the square root, to get

$$\text{Estimated st. dev.} = \sqrt{\frac{\Sigma f(m - \bar{X})^2}{n - 1}}$$

$$= \sqrt{5248/49} = 10.34901.$$

To estimate the quartiles, GROUP identifies the quartile classes, using the cumulative frequency. With n = 50 observations the first quartile will cut off the lowest n/4 observations, or n/4 = 50/4 = 12.5. We therefore assume that Q1 lies in the third class "20 but less than 30." Since the lowest six observations are already included in the first two classes (0 < 10, 10 < 20) and there are 23 observations in the third class, it is logical to interpolate within the third class, assuming that all observations within the class are uniformly distributed. To the class lower limit (20) we add a fraction of the class width:

$$\hat{Q}_1 = Q_1 \text{ class lower limit} + \left[\frac{(n/4 - \text{prior cum F})}{Q_1 \text{ class frequency}}\right] (\text{class width})$$

$$= 20 + \frac{(12.5 - 6)}{23} (10)$$

$$= 20 + 2.826$$

$$= 22.826.$$

Formulas for estimating the other quartiles are similar. The median (second quartile) divides the fifty sorted observations into two equal groups of twenty-five (since n/2 = 50/2 = 25) to define the top and bottom half of the sorted array. The median is therefore somewhere within the third class, "20 but less than 30." GROUP uses n/2 in the formula to interpolate within the median class, "20 but less than 30":

$$\hat{Q}_2 = Q_2 \text{ class lower limit} + \left[\frac{(n/2 - \text{prior cum F})}{Q_2 \text{ class frequency}} \right] (\text{class width})$$

$$= 20 + \frac{(25 - 6)}{23} (10)$$

$$= 20 + 8.261$$

$$= 28.261.$$

The third quartile separates the lowest 75 percent and highest 25 percent of the observations, so it must lie in the fourth class, "30 but less than 40" since 3n/4 = 3(50)/4 = 37.5. GROUP uses this formula to interpolate:

$$\hat{Q}_3 = Q_3 \text{ class lower limit} + \left[\frac{(3n/4 - \text{prior cum F})}{Q_3 \text{ class frequency}} \right] (\text{class width})$$

$$= 30 + \frac{(37.5 - 29)}{13} (10)$$

$$= 30 + 6.538$$

$$= 36.538.$$

The three quartiles should be assumed to lie in different classes. However, they could be in the same class. In this case the first and second quartiles happen to lie within the same class, "20 < 30." GROUP will highlight the median class. You may identify the other quartile classes by examining the cumulative frequencies (last column in the GROUP printout).

How accurate are the grouped estimates of the mean, standard deviation, and quartiles? Usually, we have no way of knowing. But suppose we happened to find the source of the raw data, as shown in Table 12.2. Using ANALYZ (or BOXPLOT), we could calculate the same statistics from the raw data to compare with GROUP's estimates.

In this case, GROUP's estimates are reasonably good, as shown in Table 12.3. (Estimates are rounded for easier comparison.) We find that GROUP's estimated first quartile is too low, while the estimated third quartile is too high. GROUP's other statistics are fairly close to the "true" (ungrouped) statistics from ANALYZ. Of course, if we had the raw data, we could have used ANALYZ in the first place.

Accuracy of grouped estimates will be affected by the distribution of data within each class. Usually, we have no way of knowing whether the data are uniformly distributed within each class. However, if raw data were available, we could check. For example, a more detailed histogram for the modal third class, "20 < 30," looks like Figure 12.1.

Table 12.2   Performance of Selected Growth-Oriented Mutual Funds

Fund	Return	Fund	Return
ABT Utility Inc	18.1	Mass Cap Dev	42.9
Alliance Surveyor	28.4	Merrill Lynch Bas Val	27.7
Am Cap Enterprise	26.9	Merrill Lynch Spc Val	26.6
Am New Economy	21.1	Mutual Shares Corp	27.8
BLC Growth	28.7	National Aviat & Tech	19.1
Bullock Canadian	4.4	NEL Growth	33.5
Charter Fund	32.1	Nicholas Fund	25.5
Copley Tax Managed	32.8	Oppenheimer Regency	17.9
Dean-Witter Ind Val	36.4	Penn Mutual	26.7
Dreyfus Third Cent	24.7	Pro Fund	32.3
Euro Pacific Growth	40.7	Pru-Bache Research	43.8
Fidelity Capital Gro	32.3	Putnam Investors	24.4
Fidelity Discoverer	31.5	Reich & Tang Equity	39.8
Fidelity Magellan	52.0	Scudder Cap Growth	37.5
Fidelity Selct — Dfns	19.9	Seligman Comm & Info	27.2
Fidelity Selct — Leis	52.2	Sequoia Fund	25.9
Franklin Equity	30.8	Sigma Special	25.0
Fund of the Southwst	24.0	Sigma Venture Shares	27.2
GT Pacific	33.7	Stein Roe Universe	35.5
IDS Growth	31.3	T Rowe Price New Era	28.5
Investors Research	25.8	Templeton Growth	27.2
JP Growth	24.0	Twentieth Cen Growth	45.8
Keystone Intl	48.2	United Accumulative	22.0
Lehman Capital	27.5	USAA Cornerstone	24.3
Lindner Fund	19.9	Vanguard Sp Ptf—Hth	45.5

Caution: These data are intended solely for classroom purposes and should not be viewed as a guide to investments.
Source: *Money*, Vol. 15, No. 5, May, 1986, pp. 208–234.

Table 12.3   Comparison of Ungrouped and Grouped Estimates

Statistic	Ungrouped (ANALYZ)	Grouped (GROUP)
Mean	30.1	29.8
St. Dev.	9.3	10.4
First quartile	24.7	22.8
Second quartile	27.8	28.3
Third quartile	33.7	36.5

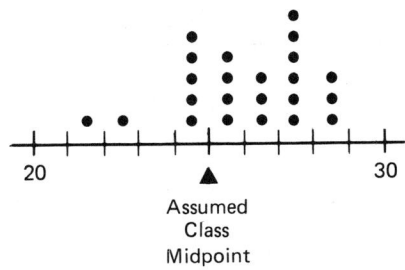

Figure 12.1   Dot Diagram for Modal Class

Figure 12.1 shows that the actual class mean is 25.9 (not 25), which will have a slight adverse effect on GROUP's estimate of the mean. Also, interpolation within the class will be imprecise because a majority of the observations (16 out of 23) are above the class midpoint. When data are Gaussian (normal) the simplifying assumption of a uniform distribution within classes is weak.

Whether GROUP's estimates are seriously damaged depends on the other classes. Skewed observations within one class may be offset by other classes. Unless there is a systematic skewness within every class (say, clustering at the low end of each class), the biases will tend to average out and nonuniform data in one class will not ruin GROUP's estimates.

The more classes we have, the better GROUP's estimates become in the sense that they will be closer to the true ungrouped mean, standard deviation, and quartiles. For example, if instead of six classes of width ten we had twelve classes of width five ($0 < 5$, $5 < 10$, and so on), we might have come closer to the results obtained by using ANALYZ. But we usually have no control over the number of classes—data are often presented in grouped form only (and if we had the raw data, we would probably use ANALYZ instead of GROUP).

Conversely, if we were to collapse our classes to only three classes of width twenty ($0 < 20$, $20 < 40$, $40 < 60$), GROUP's estimates would be less accurate, since more information would have been lost. Trial-and-error may convince you that decent ball park estimates can usually be made even from only a few classes. The ideal number of classes for grouped data might be sought by using Sturges's Rule (see Chapter 3) if you wish to be formal about it.

In real life we must take whatever is given by the author of the grouped data table. Unfortunately, open-end top classes are quite common (such as ''250,000 and over''). Open-end classes prevent estimation of the mean and standard deviation, since there is no midpoint for an open-end class. But quartiles often can be obtained, depending on the data. Although GROUP cannot officially handle such data, you could put in a false but realistic upper class limit and then ignore the mean and standard deviation. So long as the quartiles do not fall within the open-end class, they should be acceptable, giving you at least a general idea of central tendency and dispersion.

Open-end bottom classes (such as ''250 and below'') are rare. The usual motive for open-end classes is unusually large values rather than unusually small ones. If you do have an open-end bottom class, you may be able to assume a logical lower limit (usually zero). Otherwise, try the approach suggested above for open-end top classes to force GROUP to yield at least some statistics.

◼ Case Study 12.2: Heating Degree-Days

A sample of U.S. cities is taken by selecting every other city from a list of U.S. weather-reporting stations. For each city, degree-days are recorded for the month of December. (A degree-day is defined as the sum over all days in the month of the difference between 65 degrees Fahrenheit and the daily mean temperature for each day.) The results are presented in grouped form in Table 12.4.

Since the classes are unequal, we must use GROUP's option to specify individual class limits. The data input is quick, since there are only four classes.

It might be expected that the estimates from GROUP would be poor, since only four classes are used and there is much clustering in the last two classes. Suppose we later discover the raw data, shown in Table 12.5, which can be used to produce ungrouped estimates (using ANALYZ).

Probably the most serious problem for GROUP is that the midpoint of our last class, $1000 < 2000$, gives a very poor idea of the location of data within that class. Only two cities (Duluth at 1587 and Bismarck at 1538) are even close to the class midpoint of 1500, while the rest of the cities are between 1000 and 1300. This will cause a serious upward bias in the mean, since GROUP assumes that all 14 cities in the last class are at the class midpoint of 1500.

It can be seen from Table 12.6 that the GROUP estimates are not very good. The GROUP mean and standard deviation are too high, and so is the

Table 12.4   December Heating Degree-Days in Selected Cities

Degree-Days	Frequency
0 but under 250	2
250 but under 500	5
500 but under 1000	14
1000 but under 2000	14
Total:	35

Source: U.S. Bureau of the Census, *Statistical Abstract of the United States, 1986* (Washington, D.C.: U.S. Government Printing Office, 1985), p. 219.

```
                              GROUP

This program calculates various sample statistics such as the
mean, median, standard deviation, and coefficient of variation
from grouped data.  Closed-end classes (up to 15) are assumed.
You will be asked to input the number of classes, interval
size, class lower limit(s), and frequencies for each class (up
to N=9999). Detailed calculations will be shown on the screen.

How many classes (up to 15)? 4
Equal interval size (y/n)? n
Lower limit of class 1 ? 0
Lower limit of class 2 ? 250
Lower limit of class 3 ? 500
Lower limit of class 4 ? 1000
Upper limit of class 4 ? 2000

INPUT FREQUENCIES FOR EACH CLASS INTERVAL

Class                   Frequency
-------------           ---------
      0 <    250        ? 2
    250 <    500        ? 5
    500 <   1000        ? 14
   1000 <   2000        ? 14
```

```
          SUMMARY OF CALCULATIONS USING GROUPED DATA
    ------------------------------------------------------------
                                                   _   2
       Class         f      m      f*m      f*(m-x)     Cum f
    ------------------------------------------------------------
      0 <    250      2     125      250    1396836.8      2
    250 <    500      5     375     1875    1715306.3      7
    500 <   1000     14     750    10500     621607.2     21
   1000 <   2000     14    1500    21000    4071607.0     35
    ------------------------------------------------------------
    Total:           35            33625    7805357.0
```

```
Note: Median class is highlighted

              SUMMARY STATISTICS FOR GROUPED DATA

Mean = 960.7143
St. Dev. = 479.1339 (using sample formula)
St. Dev. = 472.2396 (using population formula)
Coef. of Variation =  49.9 (using sample formula)
Coef. of Variation =  49.2 (using population formula)
1st Quartile = 562.5
2nd Quartile = 875
3rd Quartile = 1375

Where now:   E=EXPLORE menu   R=Run GROUP again   X=exit ? e
```

third quartile. Table 12.4 is not a very informative grouping (you could certainly do better if you made the classes), but it does illustrate that GROUP
can yield useful results, even with poor raw material. We might have expected
even worse, given the scant number of classes and completely unfavorable distribution of data within the four class intervals.

Table 12.5   December Heating Degree-Days in Selected Cities

City	Deg.-Days	City	Deg.-Days
Mobile	382	Concord	1256
Phoenix	368	Albuquerque	911
Los Angeles	255	Buffalo	1122
San Francisco	490	Charlotte	694
Hartford	1113	Bismarck	1538
Washington	809	Cleveland	1051
Miami	42	Oklahoma City	778
Honolulu	0	Philadelphia	915
Chicago	1156	Providence	1014
Indianapolis	1039	Sioux Falls	1404
Wichita	949	Nashville	747
New Orleans	336	El Paso	639
Baltimore	884	Salt Lake City	1076
Detroit	1132	Norfolk	667
Duluth	1587	Seattle	744
Jackson	513	Charleston	871
St. Louis	955	Cheyenne	1107
Omaha	1172		

Source: U.S. Bureau of the Census, *Statistical Abstract of the United States, 1986* (Washington, D.C.: U.S. Government Printing Office, 1985), p. 219.

Table 12.6   Comparison of Ungrouped and Grouped Estimates

Statistic	Ungrouped (ANALYZ)	Grouped (GROUP)
Mean	849.0	960.7
St. Dev.	382.6	479.1
First Quartile	639.0	562.5
Second Quartile	911.0	875.0
Third Quartile	1113.0	1375.0

■ Exercises

12.1   Run ANALYZ with a cross-sectional file from the state database. Print the key statistics and histogram. Then try another histogram using more classes and a third histogram using fewer classes. (You will have to choose the class limits yourself.)

12.2   Run GROUP on the frequencies from your three ANALYZ histograms. Print the screens.

12.3 Make a side-by-side comparison of the ungrouped statistics (ANALYZ) and grouped estimates (GROUP). How much accuracy was lost? Discuss the effect of increasing or decreasing the number of classes.

12.4 For the first histogram, find the class with the largest frequency (modal class). Manually plot the actual distribution of raw data within the class interval. What is the implication for the GROUP estimates?

12.5 "Estimates from grouped data are inherently inaccurate. My GROUP estimates are way off!" lamented Paula Puorg, a disgruntled statistics student. Discuss what might have thrown her estimates off.

12.6 Run GROUP with the data in Table 12.7. How accurate do you believe your grouped estimates will be? Why?

Table 12.7 Annual Population Increase in World Nations

Percent Increase	Number of Nations
5 but less than 6	1
4 but less than 5	10
3 but less than 4	30
2 but less than 3	50
1 but less than 2	31
0 but less than 1	27
−1 but less than 0	8
Total:	157

Source: U.S. Bureau of the Census, *Statistical Abstract of the United States, 1986* (Washington, D.C.: U.S. Government Printing Office, 1986), pp. 835–837.

12.7 Run GROUP with the data in Table 12.8.

Table 12.8 Life Expectancy at Birth in Large Nations

Life Span (Years)	Number of Nations
70 but less than 80	20
60 but less than 70	12
50 but less than 60	13
40 but less than 50	9
30 but less than 40	2
Total:	56

Source: U.S. Bureau of the Census, *Statistical Abstract of the United States, 1986* (Washington, D.C.: U.S. Government Printing Office, 1986), p. 839. Includes only countries with populations of ten million or more.

Suppose a finer breakdown of frequencies within the interval "70 but less than 80" becomes available after you have run GROUP:

Age	f	Age	f
79	0	74	3
78	0	73	3
77	0	72	2
76	2	71	4
75	3	70	3

What effect, if any, would this finer breakdown have on your assessment of the accuracy of GROUP's estimates?

12.8 Which sample statistics, if any, can be obtained from the data in Table 12.9? Try to run GROUP with these data. Discuss the problems that arise. Why were unequal class intervals used in this table? Try to rewrite the class limits using equal widths, and discuss the problems that arise.

Table 12.9   Size of U.S. Farms, 1982

Size (acres)	Farms (thousands)
Under 10	188
10 but less than 50	449
50 but less than 100	344
100 but less than 180	368
180 but less than 260	211
260 but less than 500	315
500 but less than 1000	204
1000 but less than 2000	97
2000 and over	65
Total:	2241

Source: U.S. Bureau of the Census, *Statistical Abstract of the United States, 1986* (Washington, D.C.: U.S. Government Printing Office, 1986), p. 636.

Optional Exercises for Ambitious Learners

12.9 What happens in Problem 12.8 if you run GROUP using the actual frequencies (that is, multiplying the given frequencies by 1000 to get 188000, 449000, and so on)? Should you? If not, why not? What difference does it make?

12.10 Which sample statistics, if any, can be obtained from the data in Table 12.10? Run GROUP with these data, and discuss the problems you encounter. Then express the frequencies in millions, to the nearest whole million (17, 24, 16, 10, 5, 5) and rerun GROUP. Discuss the results. What happens if you try to run GROUP after multiplying the class limits by 1000 (to get 10000, 20000, and so on) and express frequencies in raw units (17,379,000, 23,902,000, and so on)? What advice would you give a novice about choosing units of measurement?

Table 12.10   Individual Adjusted Gross Income, 1983

Income Class (thousands)	Taxpayers (thousands)
Under 10	17,379
10 but less than 20	23,902
20 but less than 30	16,032
30 but less than 40	10,353
40 but less than 50	5,127
50 and over	5,214
Total:	78,007

Source: U.S. Bureau of the Census, *Statistical Abstract of the United States, 1986* (Washington, D.C.: U.S. Government Printing Office, 1986), p. 317.

12.11 In the limit, if you reduce the number of classes to one, what happens to a grouped data analysis? Try running GROUP with your data set from Exercise 12.1 reduced to one class of n observations. Discuss the accuracy of your result, if any. Going in the other direction, is there any logical maximum to the number of classes? Discuss.

12.12 Run GROUP with the data in Table 12.11. Discuss perceptions of the market for computers, subsystems, and peripherals as seen by companies in the lowest quartile, the average company, and the top-quartile companies. Note that an outlier was excluded from Table 12.11 (Iomega Corporation with a 423 percent revenue increase). If we had tried to include it, what problems would arise with GROUP? Can you think of ways to include it? What do you suppose explains the existence of the outlier? Are there any other extreme points? What, if anything, could be done to make the table more informative? What would Sturges's Rule say about the number of classes for this data?

Table 12.11   One-Year Revenue Change, Computer Firms

Revenue Change (percent)	Number of Firms
−80 but less than −60	1
−60 but less than −40	0
− 40 but less than −20	1
−20 but less than 0	13
0 but less than 20	17
20 but less than 40	23
40 but less than 60	9
60 but less than 80	7
80 but less than 100	0
100 but less than 120	3
120 but less than 140	3
140 but less than 160	2
Total:	79

Source: Sumner N. Levine, *1986 Dow Jones—Irwin Business and Investment Almanac* (Homewood, Illinois: Dow Jones—Irwin Publishing Company, 1986), pp. 72–73. One outlier (Iomega Corporation at 423 percent increase) was omitted.

12.13 Run GROUP on the data shown in Table 12.12. Then refer back to the ungrouped data in Exercise 11.1 (see Table 11.7). Check the accuracy of GROUP's estimates and discuss. Is Table 12.12 a "good" classification or not?

Table 12.12   Years Served by Roman Catholic Popes Since 1500

Years Served	Number of Popes
0 but less than 5	14
5 but less than 10	12
10 but less than 20	17
20 but less than 40	6
Total:	49

Source: *Reader's Digest 1982 Almanac* (Pleasantville, N.Y.: Reader's Digest Association, 1982), p. 709.

13

LAYOUT/Displaying
Data Files

Introduction

This program will read several data files and print their contents in parallel columns. Each file should represent a different variable. The printing format will be "nice," meaning that there will be column headings, a case count will be provided, and decimal points will all be lined up. Integer data will be right-justified, while character data will be left-justified. In other words, this program tries to do things the same way people would do them, so the results will look nice. To save space, LAYOUT prints only the first fifteen columns of a label file, and the remaining characters are truncated.

Files need not contain the same number of cases, but you would usually use this program to display files in a certain database, so most of the time your columns will have the same length. The program is useful for constructing appendixes for reports. For example, you might wish to list some of the state data files that you are using in a regression.

The number of files that can be listed side by side varies, depending on the data itself. The maximum is twelve files. The computer will print as much as it can and will simply omit any file that will not fit on the page. The minimum and maximum data field width depend on the column heading and data values to be printed.

◼ Case Study 13.1: Profile of Chief Executive Officers

Table 13.1 shows a sample of nineteen chief executive officers (CEOs), chosen at random from an alphabetical list of the 793 largest companies in the United States. The sampling methodology was to choose the tenth name from the top of each page in the nineteen-page table. For each CEO in our sample, we record the company name and the CEO's name, age, total compensation (thousands of dollars, including bonuses and stock gains), years of experience in the company, and type of business background.

Table 13.1   Profile of Randomly Chosen CEOs in Large U.S. Firms

Firm	Chief Executive	Age	Compen- sation	Years at Firm	Back- ground
Alexander & Baldwin	R.J. Pfeiffer	66	1327	26	Operations
AmeriTrust	J.V. Jarrett	54	771	12	Banking
Bell Atlantic	T.E. Bolger	58	909	43	Admin.
Centex	P.R. Seegers	56	466	24	Finance
Comdisco	K.N. Pontikes	46	1409	16	Marketing
Digital Equip.	K.H. Olsen	60	755	28	Founder
Fin. Corp. Santa Barb.	P.R. Brinker- hoff	43	808	1	Finance
Fourth Financial	J.L. Haines	59	297	24	Admin.
GTE	T.F. Brophy	63	1056	28	Legal
IBM	J.F. Akers	51	736	25	Marketing
Louisiana Banc- shares	C.W. McCoy	66	220	26	Banking
Mesa Petroleum	T.B. Pickens, Jr.	58	9877	29	Founder
N. Amer. Philips	C. Bruynes	53	864	11	Marketing
Pfizer	E.T. Pratt, Jr.	59	2260	21	Finance
Roadway Services	C.F. Zodrow	63	508	27	Finance
Squibb	R.M. Furlaud	63	2780	30	Legal
Thrifty Corp.	L.H. Straus	71	1118	40	Legal
USF & G	J. Moseley	55	420	32	Insurance
Yellow Freight Sys.	G.E. Powell, Jr.	60	662	34	Admin.

Compensation is in thousands of dollars.
Source: *Forbes,* Vol. 137, No. 12, June 2, 1986, pp. 152–188.

These data will be used in another case study (see Chapter 21) and are used here only to illustrate how to obtain nicely formatted column data to use in your reports. Using the EXPLORE file editor, we create six files: CEO, FIRMNAME, BACKGR, COMPENS, EXPRNC, and AGES. Then we use LAYOUT to print them as shown in the printout at the end of this chapter. Note that in the printout the last few characters of files FIRMNAME and CEO are cut off to save space. Even with these space-saving features, the file AGES will not fit on the eighty-column screen. Sensing this problem, LAYOUT suppresses printing AGES.

LAYOUT

This program uses automatic formatting to print
several disk data files simultaneously in nice columns.
Maximum file length is 500 items. Maximum number of
files that fit side-by-side depends on format chosen
by the computer (theoretically, up to 12 files). Go
ahead and try to print as many files as you want -- the
last few will be omitted if they won't fit. Size of
print fields depends on data. Files may contain non-
numeric data, but column width maybe truncated. Your
first file could be case labels such as STATE or YEAR.

FILE INPUT:

How many files to print side-by-side? 6

Name of file 1 ? firmname
Name of file 2 ? ceo
Name of file 3 ? backgr
Name of file 4 ? compens
Name of file 5 ? exprnc
Name of file 6 ? ages

Please wait ... setting print fields:

FIRMNAME : Alphabetic data
CEO   : Alphabetic data
BACKGR   : Alphabetic data
COMPENS : Numeric data
EXPRNC   : Numeric data
AGES   : Numeric data

Case	FIRMNAME	CEO	BACKGR	COMPENS	EXPRNC
1	Alexander & Ba	R. J. Pfeiffer	Operations	1327	26
2	AmeriTrust	J. V. Jarrett	Banking	771	12
3	Bell Atlantic	T. E. Bolger	Admin	909	43
4	Centex	P. R. Seegers	Finance	466	24
5	Comdisco	K. N. Pontikes	Marketing	1409	16
6	Digital Equip	K. H. Olsen	Founder	755	28
7	Fin Corp Santa	P. R. Brinkerh	Finance	808	1
8	Fourth Financi	J. L. Haines	Admin	297	24
9	GTE	T. F. Brophy	Legal	1056	28
10	IBM	J. F. Akers	Marketing	736	25
11	Louisiana Banc	C. W. McCoy	Banking	220	26
12	Mesa Petroleum	T. B. Pickens	Founder	9877	29
13	N Amer Philips	C. Bruynes	Marketing	864	11
14	Pfizer	E. T. Pratt Jr	Finance	2260	21
15	Roadway Servic	C. F. Zodrow	Finance	508	27
16	Squibb	R. M. Furlaud	Legal	2780	30
17	Thrifty Corp	L. H. Straus	Legal	1118	40
18	USF & G	J. Moseley	Insurance	420	32
19	Yellow Freight	G. E. Powell J	Admin	662	34

Where now: E=EXPLORE menu   R=Run LAYOUT again   X=exit ? e

One minor caution: Commas in a label file can cause errors. For this reason the commas were omitted from the file CEO. This problem arises from the way BASIC handles print formatting. (If you can find a way around it, let the author know.)

■ **Exercises**

13.1 Choose any group of the state data files from the DATABASE-1 diskette and list them side by side, using LAYOUT. Use the file STATE in the first column to label the other files. Refer to Appendix H if necessary.

13.2 Choose any group of the time-series data files from the DATABASE-1 diskette and list them side by side, using LAYOUT. Use the file YEAR in the first column to label the other files. Refer to Appendix I if necessary.

13.3 Take a close look at Appendixes H and I. You can generate tables approximately like them using LAYOUT. Remember that LAYOUT can help you generate an appendix for your statistical reports by listing files of interest. The first two exercises are intended to give you a little advance practice using LAYOUT for other exercises, such as those involving regression (program MGRES) or the chi-square test for independence (program CHI). LAYOUT comes in handy any time you want to keep track of what is in your files. Notice how important it is to list a label variable in column 1.

14

MATCOR/Matrix of Correlation Coefficients

Introduction

Sometimes, all we really need to know about a group of variables is which ones are correlated with which others. That is what MATCOR provides: a compact matrix of simple correlation coefficients. The maximum number of variables is fifteen, with no more than 200 observations per variable. Each variable must be saved in a data file. If more than seven data files are used, the matrix will be printed in partitioned form (each portion of the matrix will be tailored to fit on one screen).

All that is required of the data is that the level of measurement be at least interval (that is, not nominal or ordinal except when a variable is coded as a binary). For most economic or accounting data this condition is easily met. It is at least arguably met by using survey data with Lickert-scaled responses, such as might be obtained in marketing (for example, perceptions of competing products) or organizational behavior (for example, attitudes toward a job).

It should be noted that MGRES includes a matrix of correlation coefficients as part of the printout, so if you are doing a regression, you need not run MATCOR too.

MATCOR contains diagnostic messages for occasions when you try to compare files of different length. Naturally, you cannot compare files from different databases, since the idea of a bivariate correlation coefficient presumes n *pairs* of observed data (X_i, Y_i). Unpaired data cannot be compared.

■ Case Study 14.1: Financial Sector Interrelatedness

Economists have long been aware that most economic time-series variables are interrelated. In many cases there is a true cause-and-effect relationship. However, it is rarely anything as simple as a bivariate relationship. Usually, rather elaborate models must be proposed to interpret real-world economic data.

To illustrate the complex interrelatedness of almost all economic time se-

Table 14.1 Selected U.S. Financial Statistics, 1960–1985

Year	Prime Rate	Dow-Jones Industrial Average	Money Supply Components			Federal Debt
			M1	M2	M3	
1960	4.82	618.04	141.8	312.3	315.3	290.9
1961	4.50	691.55	146.5	335.5	341.0	292.9
1962	4.50	639.76	149.2	362.7	371.4	303.3
1963	4.50	714.81	154.7	393.2	406.0	310.8
1964	4.50	834.05	161.9	424.8	442.5	316.8
1965	4.54	910.88	169.5	459.4	482.2	323.2
1966	5.63	873.60	173.7	480.0	505.1	329.5
1967	5.61	879.12	185.1	524.3	557.1	341.3
1968	6.30	906.00	199.4	566.3	606.2	369.8
1969	7.96	876.72	205.8	589.5	615.0	367.1
1970	7.91	753.19	216.6	628.2	677.5	382.6
1971	5.72	884.76	230.8	712.8	776.2	409.5
1972	5.25	950.71	252.0	805.2	886.0	437.3
1973	8.03	923.88	265.9	861.0	985.0	468.4
1974	10.81	759.37	277.5	908.4	1070.4	486.2
1975	7.86	802.49	291.1	1023.1	1172.2	544.1
1976	6.84	974.92	310.3	1163.6	1311.8	631.9
1977	6.83	894.63	335.3	1286.6	1472.5	709.1
1978	9.06	820.23	363.0	1388.9	1646.4	780.4
1979	12.67	844.40	389.0	1497.9	1803.6	833.8
1980	15.27	891.41	414.8	1631.4	1988.5	914.3
1981	18.87	932.92	441.8	1794.4	2235.8	1003.9
1982	14.86	884.36	480.8	1954.9	2446.8	1147.0
1983	10.79	1190.34	528.0	2188.8	2701.7	1381.9
1984	12.04	1178.48	558.5	2371.7	2995.0	1576.7
1985	9.93	1328.23	624.7	2563.6	3213.5	1827.5

Dollar figures are in billions of current dollars.
Source: *Economic Report of the President* (Washington, D.C.: U.S. Government Printing Office, 1986).

ries, consider the matrix of U.S. macroeconomic financial data shown in Table 14.1. These variables are chosen from the time-series data files on the DATA-BASE-1 diskette. The variables are chosen without any particular model in mind, merely to show how strong some of the correlations really are.

The correlation coefficient (denoted r) ranges from +1 (signifying a perfect positive correlation) to −1 (signifying a perfect inverse correlation). When r is near zero, it signifies the absence of any bivariate statistical association between X and Y. MATCOR's formula for the correlation coefficient is

$$r = \frac{\Sigma(X_i - \overline{X})(Y_i - \overline{Y})}{\sqrt{\Sigma(X_i - \overline{X})^2}\sqrt{\Sigma(Y_i - \overline{Y})^2}}.$$

A commonly used quick rule of thumb is that a bivariate correlation is *probably significant* if

$$|r| > 2/\sqrt{n}.$$

Since our sample size is n = 26 years, this rule would say that a correlation between any two financial variables is probably significant if

$$|r| > 2/\sqrt{26}$$
$$|r| > .3922$$

This rule of thumb is derived from a simplified two-tail Student's t test with n − 2 degrees of freedom at the .05 level of significance. The rule works fairly well, especially in larger samples.

MATCOR (see sample printout) shows that according to our rule of thumb, *all but one* of these financial variables are significantly intercorrelated! While they are, in fact, related in some complex causal way, it would be unwise to draw inferences about cause-and-effect from any *single* correlation coefficient, for three reasons.

First, a large part of the observed correlation is caused by trend alone. For example, it is easy to see that the federal debt (FDEBT) and "high-powered money" (M1) have strong upward trends over time. It would be desirable to remove this trend (using a program like TREND) before drawing any strong conclusions. While FDEBT and M1 may be causally related, it is quite possible for causally *unrelated* variables to exhibit spurious correlation due to trend alone.

Second, relationships in economic or financial data are seldom bivariate in nature. It is usually best to attempt to build a complete model that *simultaneously* incorporates all relevant variables, using a technique such as multiple regression. This suggestion holds also for most data encountered in marketing and behavioral research.

Third, correlation demonstrates only *association,* not *causation.* While there are undoubtedly cause-and-effect connections among these variables, the presence of a statistical association does not *per se* prove anything except that there exist promising lines of inquiry for statistical model-building.

Despite these cautions, a correlation matrix is an essential part of many investigations. This case study is intended to suggest a sense of caution about excessive generalizations from bivariate relationships without discouraging the use of MATCOR. Since high correlations are easy to find in financial and economic data, economists consider it unsporting to make much of them, which is the main point of this case study. In marketing or behavioral data, high correlations are rare and could cause more excitement.

```
                        MATCOR

        This program will read several files, and will print
a compact matrix of simple correlation coefficients.  Each
file is assumed to contain N observations on one variable.
Files must, of course, be of equal length.  It's best to
use short file names (8 characters or less) since the format
is compressed, and long file names may result in strange
heading spacing.  Maximum is 17 variables and 400 cases.

How many files to read? 6
Name of file # 1 ? prime
Name of file # 2 ? djia
Name of file # 3 ? m1
Name of file # 4 ? m2
Name of file # 5 ? m3
Name of file # 6 ? fdebt


            TABLE OF MEANS AND STANDARD DEVIATIONS
    ====================================================
    Variable        Mean         St. Dev.      Cases
    ----------------------------------------------------
    PRIME          8.292307      3.874613        26
    DJIA           883.0328      159.9508        26
    M1             294.9115      140.7506        26
    M2             1047.25       680.6876        26
    M3             1231.719      884.0204        26
    FDEBT          645.3923      428.4316        26
    ----------------------------------------------------
    Note: In this table, the standard deviation uses (n-1)


              MATRIX OF CORRELATION COEFFICIENTS

              PRIME     DJIA     M1       M2       M3      FDEBT

    PRIME     1.000    0.366    0.762    0.763    0.769    0.675
    DJIA      0.366    1.000    0.779    0.773    0.768    0.806
    M1        0.762    0.779    1.000    0.999    0.998    0.980
    M2        0.763    0.773    0.999    1.000    0.999    0.980
    M3        0.769    0.768    0.998    0.999    1.000    0.983
    FDEBT     0.675    0.806    0.980    0.980    0.983    1.000

    Where now:  E=EXPLORE menu    R=Run MATCOR again    X=exit ? e
```

◼ Exercises

14.1 From the DATABASE-1 diskette, choose several state data files that you think might be related. Run MATCOR to find out what the correlations look like. Use the quick rule to decide which pairs of variables appear related and which appear to be unrelated. Summarize your conclusions in the form of a table designed for quick assimilation.

14.2 Repeat Exercise 14.1 using several time-series variables from the DATA-BASE-1 diskette.

14.3 Does your experiment support the idea that time-series variables are more likely to be correlated than cross-sectional data (such as the state data files)? If so, can you explain why this situation might logically exist for the particular files you chose? If not, can you explain why not, giving specific examples from the files you chose?

14.4 Discuss how one of your chosen variables might *cause* a change in another of your chosen variables. Be specific about the chain of real-world events that might transmit the change from one variable to the other. Might there be time lags before the effects are felt? If so, how would it alter the ideal procedures for measuring the correlation?

14.5 Does the lack of a correlation between two files X and Y mean that there is no cause-and-effect relationship between them? Discuss your reasoning fully, giving simple, realistic examples to demonstrate your point(s) if possible.

14.6 Does the presence of a significant correlation between two variables X and Y mean that there is a cause-and-effect relationship between them? Discuss your reasoning fully, giving simple, realistic examples to demonstrate your point(s) if possible.

References

1. Ames, Edward and Reiter, Stanley, "Distribution of Correlation Coefficients in Economic Time Series," *Journal of the American Statistical Association,* Vol. 56, No. 295, 1961, pp. 637–656.

15

MGRES/Multiple Linear Regression

Introduction

MGRES uses *ordinary least squares* (OLS) to estimate a linear equation of the form:

$$Y = \beta_0 + \beta_1 X_1 + \beta_2 X_2 + \ldots + \beta_m X_m + \epsilon.$$

The random disturbance ϵ is assumed to have a zero mean and otherwise be well behaved. MGRES prints the estimated coefficients, t values, ANOVA table, statistics of fit, estimated Y values, and residuals (differences between actual Y and estimated Y).

MGRES also gives prominent tests for violations of the assumptions of regression: non-normal disturbances (skewed or non-bell-shaped residual histogram), autocorrelated disturbances (residual pattern over time), or heteroskedastic disturbances (nonconstant residual variance). There is an option to save the estimated Y values and residuals for further analysis.

The multivariate data set will consist of n observed values of the dependent variable (Y) and n observed values of the independent predictors (X_1, X_2, . . . , X_m). The data may be regarded as a matrix (see Figure 15.1), each column of which represents one variable. The first column (column of 1s) represents the intercept term in the model, so X_0 is not really a variable like the other predictors.

The problem is to choose the vector of m + 1 unknown coefficients (β_0, β_1, β_2, . . . , β_m) so that the sum of squared residuals for the data set is minimized. Using the computer, it is relatively easy to solve this problem. The vector of estimated coefficients that ensures this result is found by using the rules of matrix arithmetic, which involves first transposing and multiplying a couple of matrices, then carrying out a matrix inversion. In matrix notation the matrix to be inverted is $X'X$, where X is the right-hand portion of the above schematic. The OLS estimate is given by

$$\hat{\beta} = (X'X)^{-1}X'Y$$

where $\hat{\beta}$ is the vector of estimated coefficients. If you do not like matrix notation, just let MGRES worry about the math.

Dependent Variable	Independent Variables				
Y	X_0	X_1	X_2	$\ldots$	X_m
Case 1	Y_1	1	X_{11}	X_{12}	$\ldots$	X_{1m}
Case 2	Y_2	1	X_{21}	X_{22}	$\ldots$	X_{2m}
$\vdots$	$\vdots$	$\vdots$	$\vdots$	$\vdots$		$\vdots$
Case n	Y_n	1	X_{n1}	X_{n2}	$\ldots$	X_{nm}

Figure 15.1   Organization of a Multivariate Data Set

Matrix operations can be time-consuming on small computers, even though they involve only "mere" arithmetic. The matrix inversion is the trickiest part, though not necessarily the most time-consuming. The algorithm used by MGRES is called *Choleski decomposition*. If the predictors are highly correlated (a condition known as *multicollinearity*) the matrix may be difficult to invert. Mathematically, the problem means that one column of the X matrix is almost an exact multiple of some other column(s). The $X'X$ is then nearly *singular* (a singular matrix lacks an inverse).

MGRES will detect a near-singular $X'X$ matrix and will terminate all calculations if the result is in grave doubt. However, you should be aware that in any regression where multicollinearity is present, the estimated coefficients may exhibit peculiar behavior, even if MGRES is able to carry out the calculation in a mathematically correct sense. Since MGRES can invert near-singular matrices, you alone can decide whether the multicollinearity is severe enough to undermine your regression.

The most common cause of a singular matrix is the mistaken use of the same predictor twice, thereby making one column of X an exact duplicate of another column. MGRES will detect and prevent accidental use of two predictors with the same name, but you may mistakenly have copied the same data into two files with different names. If a singularity condition arises, you should first check this possibility.

In time-series data, many variables are likely to be highly correlated (correlations of .999 are not unheard of). If you are using cross-sectional data, you may wish to remove the effect of population by expressing the variables on a per-capita basis (because any two population-related predictors will be spuriously correlated with each other).

These cautions should permit you to reduce the chances of unnecessary multicollinearity. If not, you will just have to discard one or more of the offending predictors. Although this may change the model specification from what you originally intended, it may be better than the alternative. This is a matter in which you must exercise your judgment.

Case Study 15.1: Sex Discrimination in Salaries

Does Ephemeral Products, Inc., practice sex discrimination in determining employee salaries? A random sample of twenty employee records produces Table 15.1. The sample is taken by selecting every thirtieth employee file from a cabinet containing 650 employee records, starting with the tenth file. In addition to salary we write down each employee's age, years of experience, sex (coded as a binary variable: 1 = male, 0 = female), and years of college education (6 = master's degree, 4 = bachelor's degree, 2 = associate's degree, 0 = no college degree).

We begin our statistical analysis with a careful eyeball scrutiny of the data, checking for patterns or unusual cases and just getting a feel for the magnitude and range of the numbers. For example, we observe that senior employees seem to be well paid. We also notice a few unusual cases, such as young Mary

Table 15.1 Profile of Twenty Employees of Ephemeral Products

Name	Annual Salary	Age	Years of Experience	Sex	Years of College
Mary	22,100	20	0	0	6
Frieda	36,900	31	4	0	6
Alicia	60,200	44	14	0	4
Tom	22,100	20	0	1	4
Nicole	75,900	55	25	0	2
Bill	56,200	44	14	1	0
Gillian	31,900	25	5	0	6
Bob	80,200	55	25	1	2
Vivian	68,300	50	20	0	4
Cecil	86,100	60	30	1	0
Barney	54,800	40	15	1	2
Jack	91,300	64	29	1	0
Wanda	44,800	35	10	0	6
Sam	83,100	64	19	1	2
Saundra	54,800	40	15	0	4
Pete	36,900	31	4	1	4
Steve	31,900	25	5	1	6
Fred	44,800	35	10	1	2
Dick	80,200	60	21	1	2
Lee	69,900	50	20	1	2

who has a master's degree, even though she is only 20 years old. This is possible, but unusual. It would be wise to verify the data before proceeding.

We can see some *prima facie* signs of sex-based salary inequity. For example, Mary's salary of $22,100 is identical to Tom's, even though Tom has only a bachelor's degree. In other respects, this pair of employees seems identical. A possible pattern of salary differential seems to exist for other pairs of employees, such as Alicia and Bill, Nicole and Bob, and so on.

There are twelve male employees and eight female employees. Assuming that the population proportions are similar (which should be true for a random sample), it could be said that Ephemeral Products hires about 40 percent women. Using a calculator, we find that the male mean salary is $61,458 versus $49,363 for females.

Although suggestive, this difference in mean salaries does not *per se* prove anything, since it neglects differences in education and experience. A visual inspection of the sample data gives the impression that females do have less experience. However, they seem to have more education. Since we ought to consider all salary predictors *simultaneously,* for further insight we must turn to the proposed regression model:

$$\text{SALARY} = f(\text{AGE}, \text{EXP}, \text{SEX}, \text{EDUC})$$

For each variable we use short abbreviations corresponding to the data file names that we enter using the EXPLORE file editor:

$$\begin{aligned}
\text{NAME} &= \text{employee's name} \\
\text{SALARY} &= \text{employee's annual salary (in dollars)} \\
\text{AGE} &= \text{employee's age (in years)} \\
\text{EXP} &= \text{employee's experience at Ephemeral (in years)} \\
\text{SEX} &= \text{employee's sex (1 = male, 0 = female)} \\
\text{EDUC} &= \text{employee's college education (in years).}
\end{aligned}$$

Note that we include NAME as a label file for our regression program. Details of data entry are not shown. Before proceeding, we should check our data files for possible errors (statistical consultants try to discover their errors early). One good way to begin is to run LAYOUT to display the data files side by side. If everything looks okay, we will proceed.

In linear form, with a random disturbance ϵ added at the end, the model would be

$$\text{SALARY} = \beta_0 + \beta_1 \text{ AGE} + \beta_2 \text{ EXP} + \beta_3 \text{ SEX} + \beta_4 \text{ EDUC} + \epsilon.$$

If sex discrimination exists, it will be revealed as a nonzero value for β_3. Moreover, this estimated coefficient must be *significantly* different from zero.

A priori we would expect that β_1 should be positive. Other things being equal, older employees would have higher salaries (unless age discrimination exists, in which case beyond a certain age the opposite effect would exist, making the overall effect indeterminate).

A priori we would expect that β_2 should be positive. Other things being equal, employees with greater experience would be paid more. This effect may be confounded with AGE, since as AGE increases, so does experience. However, the variables are not identical.

A priori we would expect that β_3 should be positive if there is sex discrimination in favor of males. If β_3 is zero, there is no sex discrimination; if β_3 is negative, there is reverse discrimination.

A priori we would expect that β_4 should be positive. Other things being equal, more college education should increase the employee's value to the company and remove barriers to higher salary levels.

Although we anticipate positive signs on all estimated coefficients, we are not committed to these expectations. We only want to be sure that we have thought things through carefully in causal terms before looking at the regression itself.

Before running MGRES, we run PLOT to look at the relationship between SALARY and each of the four predictors. The main results are shown in Figure 15.2. The plot of SALARY against AGE reveals a nearly perfect linear fit. The positive correlation between SALARY and EXP is also high, though perhaps not quite so nearly a straight line.

The scatter plot of SALARY against SEX tells us little, since SEX is a binary variable (0 or 1). However, we can see that male salaries (on the right) have a higher mean and are more dispersed than female salaries (on the left).

The plot of SALARY against EDUC reveals a surprising inverse relationship. An inexperienced observer might be fooled into thinking that education is penalized at Ephemeral Products, Inc. However, this paradox probably arises because older, well-paid, more experienced employees have less education than younger, lower-paid, less-experienced employees.

Bivariate scatter plots do not really tell much about what to expect from regressions. In fact, bivariate scatter plots are often misleading because they fail to control for other factors. Specifically, lack of bivariate correlation between Y and a given predictor does *not* prevent that predictor from being significant in a multivariate model. Conversely, strong bivariate correlation between Y and a given predictor does not guarantee that the predictor will be as strong in a multivariate model. In short, if the true relationship is multivariate, bivariate glimpses of the relationship are *misspecified modeling*. However, bivariate plots can help identify outliers or patterns in the data. We seem to have no distinct or unusual data points.

Next, we run MGRES. The computer requests the sample size and number of variables. This defines the size of the matrix to be manipulated (20×5). Then MGRES asks for the file names for our regression and reads the data files. We use a label file NAME to identify each employee.

The first output from MGRES consists of a table of means and standard deviations. While the computer inverts the matrix, we carefully look at the means and standard deviations to see whether they appear reasonable (we always watch for errors in reading the files). Everything looks fine so far, since the means and standard deviations are reasonable.

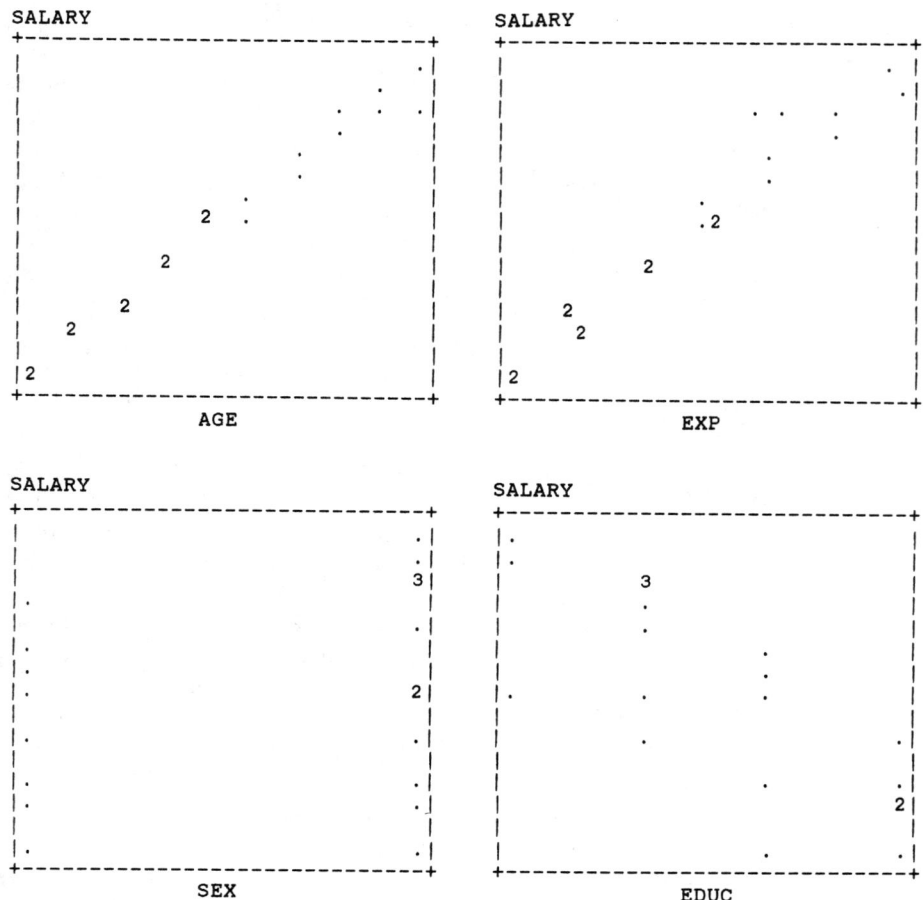

Figure 15.2   Plot of Salary Against Four Predictors

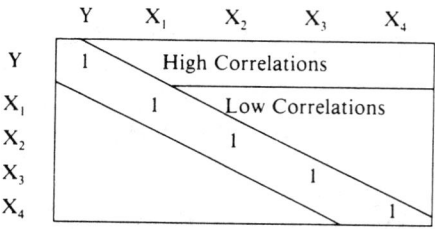

Figure 15.3   Ideal Correlation Matrix for a Regression

```
                        MGRES

        This program performs linear multiple regression
   by the Ordinary Least Squares (OLS) method.  Matrix is
   inverted by Choleski decomposition.  The model is:

   Y = B(0) + B(1)*X(1) + B(2)*X(2) + ... + B(m)*X(m)

   Each variable will be read from a separate file as a
   column of numbers.  Each file must contain N observa-
   tions, where N is the sample size.  Each file must be
   saved before running the program.  Short file names are
   desirable (8 characters or less) to ensure reasonable
   spacing in correlation matrix column headings.  You
   have the option of using a file of N case labels,
   which will be truncated to fit the assigned space. The
   maximum is 400 cases and 16 variables (including Y).

   How many observations? 20
   How many variables (including Y)? 5

   FILE NAMES FOR REGRESSION

   Which file is the dependent variable? salary
   File name for independent variable # 1 ? age
   File name for independent variable # 2 ? exp
   File name for independent variable # 3 ? sex
   File name for independent variable # 4 ? educ
   Do you want to use a label file (y/n)? y
   Name of label file? name


   TABLE OF MEANS AND STANDARD DEVIATIONS
   ========================================
   Variable        Mean           St.Dev.
   ----------------------------------------
   SALARY          56620          22066.54
   AGE             42.4           14.47466
   EXP             14.25          9.335135
   SEX             .6             .5026246
   EDUC            3.2            2.092593
   ----------------------------------------
```

In stylized form we hope to see something like the ideal correlation matrix shown in Figure 15.3. In the actual correlation matrix we could circle all correlations that exceed $2/\sqrt{n} = 2/\sqrt{20} = .447$. This quick rule of thumb suffices to identify correlations that *probably* are significant. In apparent support of our proposed hypothesis, the top row contains lots of high correlations (desirable if we want our chosen predictors to be correlated with Y).

```
              MATRIX OF CORRELATION COEFFICIENTS

                 SALARY     AGE      EXP      SEX      EDUC

       SALARY    1.0000   0.9936   0.9728   0.2755   -0.7649
       AGE       0.9936   1.0000   0.9457   0.2836   -0.7639
       EXP       0.9728   0.9457   1.0000   0.2356   -0.7706
       SEX       0.2755   0.2836   0.2356   1.0000   -0.6205
       EDUC     -0.7649  -0.7639  -0.7706  -0.6205    1.0000
```

But there is an ominously high positive correlation (.9457) between AGE and EXP (undesirable because we want our predictors to be uncorrelated with one another). Not only that, but EDUC is correlated (inversely) with every other predictor! To reduce multicollinearity, we might end up having to drop either AGE or EXP. Yet multicollinearity may not be severe, since no pairwise predictor correlation exceeds the overall multiple correlation coefficient of .9991 (this rule of thumb is known as *Klein's Rule*).

```
                A N O V A    T A B L E
===========================================================
Source of          Sum of
Variation          Squares        D.F.       Mean Square
-----------------------------------------------------------
Regression      9.235203E+09       4         2.308801E+09
Error           1.650945E+07      15         1100630
Total           9.251712E+09      19
-----------------------------------------------------------

F for analysis of variance = 2097.71
D.F. =  4 and  15 (for F-test)
R-Squared = 0.9982
Per cent of variation explained = 99.82
Multiple correlation coefficient = 0.9991
Standard error of estimate = 1049.109
```

The ANOVA table contains various measures of overall fit. This table is based on the dichotomy of variation about the mean salary:

$$\begin{array}{ccccc} \text{total} & = & \text{explained} & + & \text{unexplained} \\ \text{variation} & & \text{variation} & & \text{variation} \\ & & \text{(regression)} & & \text{(error)} \end{array}$$

or

$$\text{SST} = \text{SSR} + \text{SSE}.$$

An overall test of the model's fit is provided by the F statistic, which is a ratio of *mean square errors*. F may be calculated as

$$F = \frac{\text{SSR}/m}{\text{SSE}/(n - m - 1)}.$$

where m is the number of predictors in the model (m = 4 in our example). If F is far above unity, the regression has explanatory power. We set up an F test of the hypotheses:

H_0: $\beta_1 = 0$ and $\beta_2 = 0$ and $\beta_3 = 0$ and $\beta_4 = 0$ (regression is not significant).

H_1: Not all the βs are 0 (regression is significant).

If we use Appendix E with d.f. = (4, 15) at the .05 level of significance, the decision rule looks like Figure 15.4. Since the test statistic from the printout (F = 2097.71) is far beyond the critical value (F = 3.06), it is clear that

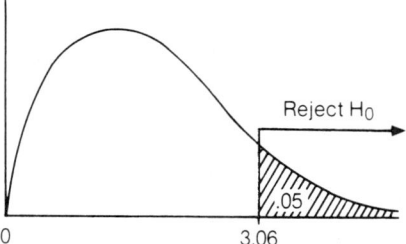

Figure 15.4   F Test for Overall Fit

the regression is significant overall (we reject H_0). One usually finds much lower F values in cross-sectional data.

Since the high r^2 suggests that the model explains 99.82 percent of the variation, at first glance there does not seem to be much remaining unexplained variation. But the standard error of the estimate (denoted SE for convenience) suggests that there is still room for improvement. MGRES calculates SE as follows:

$$SE = \sqrt{\frac{\Sigma(Y_i - \hat{Y}_i)^2}{n - m - 1}} = 1049.109.$$

In this formula, m is the number of predictors (m = 4 in our example). The smaller the standard error, the better the fit. The standard error is measured in the same units as Y and can be used to construct a confidence interval for a predicted Y value. The form of an approximate confidence interval is

$$\hat{Y} \pm tSE.$$

This approximate interval is reasonably valid if sample size is not too small and if all predictors are close to their respective means. For illustration we could use a quick rule (which assumes that t = 2) to construct an approximate 95-percent confidence interval for Y:

$$\hat{Y} \pm 2SE$$
$$\hat{Y} \pm 2(1049)$$
$$\hat{Y} \pm 2098.$$

Therefore individual salary predictions cover an interval of approximately plus or minus $2098. Stated in these terms, the model's predictive accuracy seems less impressive (though much smaller intervals would be obtained if we wanted to predict only the mean salary). A "refined" 95-percent confidence interval using the correct t value would not change this interval very much. Using d.f. = n − m − 1 = 20 − 4 − 1 = 15, Appendix C gives t = 2.131 so that:

$$\hat{Y} \pm tSE$$
$$\hat{Y} \pm 2.131(1049)$$
$$\hat{Y} \pm 2235.$$

```
                  TABLE OF ESTIMATED COEFFICIENTS
==============================================================
                Estimated         Estimated        Computed
   Variable      Coefficient       St. Dev.         t-Value
--------------------------------------------------------------
   AGE           1064.0350          52.2042          20.382
   EXP            822.1411          86.8586           9.465
   SEX           1295.6311         702.1874           1.845
   EDUC           575.3353         256.0685           2.247
   Intercept    -2829.047
--------------------------------------------------------------
```

Next, we look at the individual coefficient estimates, along with their t values. Rounding things off a bit, the estimated regression equation is

SALARY = −2829 + 1064 AGE + 822 EXP + 1296 SEX + 575 EDUC.

The negative intercept does not seem meaningful (although EXP = 0, SEX = 0, and EDUC = 0 are within the range of observed data, AGE = 0 not only is illogical in this problem but is far outside the range of observed data). All estimated coefficients are positive, as we anticipated.

On the average, each additional year of age (AGE) adds $1064 to an employee's salary. Each additional year of experience (EXP) adds $822. Being male (SEX = 1) adds $1296 to an employee's salary (which answers the sex discrimination question). Each year of college (EDUC) adds $575 to an employee's annual salary. To illustrate the sex differential, we estimate the annual salary of a hypothetical 31-year-old (AGE = 31) female employee (SEX = 0) with four years' experience (EXP = 4) and six years of college (EDUC = 6):

SALARY = −2829 + 1064 AGE + 822 EXP + 1296 SEX + 575 EDUC
$$= -2829 + 1064 (31) + 822 (4) + 1296 (0) + 575 (6)$$
$$= \$36,893.$$

An otherwise identical male employee (SEX = 1) would earn

SALARY = −2829 + 1064 AGE + 822 EXP + 1296 SEX + 575 EDUC
$$= -2829 + 1064 (31) + 822 (4) + 1296 (1) + 575 (6)$$
$$= \$38,189.$$

If regression could be construed to prove causation (which it cannot), a hypothetical sex change would be worth $1296 annually for a female employee of Ephemeral Products. Using a quick rule, we can categorize the predictors as follows:

Probably highly significant if $|t| \geq 3$;

Probably significant if $2 \leq |t| < 3$;

Probably marginally significant if $1 \leq |t| < 2$;

Probably not significant if $|t| < 1$.

This quick rule works fairly well when the sample size is not too small. By this criterion, AGE and EXP are highly significant predictors, EDUC is significant, and SEX appears to be marginally significant.

Next, we should perform formal t tests for significance on each of the four predictors (AGE, EXP, SEX, EDUC). We will illustrate with SEX, since its coefficient is of critical importance to our analysis. For the t test we use d.f. $= n - m - 1 = 20 - 4 - 1 = 15$ (where m is the number of predictors in the model). The hypotheses are

$$H_0: \quad \beta_3 = 0 \text{ (null hypothesis).}$$

$$H_1: \quad \beta_3 > 0 \text{ (suggested by sample).}$$

The test statistic for SEX from MGRES is

$$t = \frac{b_3 - 0}{\hat{\sigma}_{b_3}} = \frac{b_3}{\hat{\sigma}_{b_3}} = \frac{1295.6311}{702.1874} = 1.845.$$

The critical value of Student's t from Appendix C at the .05 level of significance is $t = 1.753$. The decision rule looks like Figure 15.5. Although we can reject H_0 at the .05 level of significance, if we had used the .01 level of significance we would *not* have rejected H_0 (critical value $t = 2.602$), as shown in Figure 15.6. In a borderline case like this a larger sample would be desirable.

It is unclear whether this evidence would stand up under courtroom scrutiny (if Ephemeral Products were charged with sex discrimination). The consistency of the pattern (comparing individual cases) might be more convincing to a judge or jury than the t values. As a refinement, it might be better to run MGRES separately for men and women (using two samples) and do away with our binary variable SEX entirely.

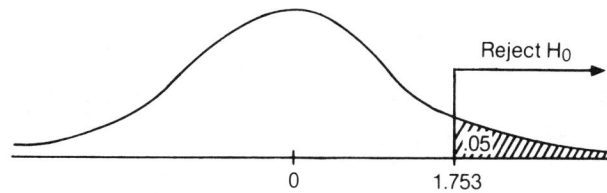

Figure 15.5   t Test for SEX Predicator at $\alpha = .05$

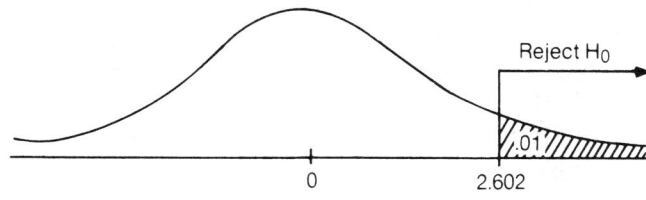

Figure 15.6   t Test for SEX Predictor at $\alpha = .01$

```
        TABLE OF ACTUAL AND ESTIMATED SALARY VALUES
========================================================
NAME          Observed       Estimated       Residual
--------------------------------------------------------
Mary           22100         21903.66          196.34
Frieda         36900         36896.61            3.39
Alicia         60200         57799.81         2400.19
Tom            22100         22048.62           51.38
Nicole         75900         77397.07        -1497.07
Bill           56200         56794.10         -594.10
Gillian        31900         31334.54          565.46
Bob            80200         78692.70         1507.30
Vivian         68300         69116.87         -816.87
Cecil          86100         86972.92         -872.92
Barney         54800         54510.77          289.23
Jack           91300         90406.91          893.09
Wanda          44800         46085.60        -1285.60
Sam            83100         83336.17         -236.17
Saundra        54800         54365.81          434.19
Pete           36900         37041.57         -141.57
Steve          31900         32630.18         -730.18
Fred           44800         45079.89         -279.89
Dick           80200         80724.32         -524.32
Lee            69900         69261.83          638.17
--------------------------------------------------------
```

Overall, it might be said that life at Ephemeral Products, Inc., is quite predictable—you get paid more just for getting older and more experienced, plus a premium for being well educated or male. These four factors explain over 99 percent of the variation in salary. The system is so deterministic that job performance can play only a limited role in salary determination. The seniority system is in full force, apparently.

From MGRES's table of estimates and residuals we identify employees whose salaries are poorly predicted by the model (large residuals). Outliers and near-outliers provide important clues about the model's performance.

We use the residuals to make a list of "overpaid" employees (large positive residual) and "underpaid" employees (large negative residual). Let's define *large* to mean "exceeding one standard error." Since the standard error is SE = 1049, we can easily circle on the MGRES table of residuals those residuals exceeding this magnitude. If we use this criterion, four employees have noteworthy salary residuals, as shown in Table 15.2.

If we could meet these people, we might find reasons why the model predicts their salaries poorly. We might find some new predictors to add to our

Table 15.2   Employees with Unusually Large Salary Residuals

"Overpaid" Employee	Positive Residual	"Underpaid" Employee	Negative Residual
Alicia	$2400	Nicole	$1497
Bob	$1507	Wanda	$1286

model to reduce the standard error even further. Maybe the employees were in different departments? had different prior work experience? different training? Maybe some employees lacked certain key managerial skills?

While it is true that the three most underpaid employees are female, remember that SEX has already been included in the model. If these three employees are underpaid, it must be due to factors *excluded* from the model. In other words, large residuals signify variation that is unexplained by our model.

Only one employee (Alicia) has an actual salary more than two standard errors higher than the MGRES estimate ($2 \times 1049 = 2098$). Alicia's actual salary is $60,200, whereas her predicted salary is $57,800. Alicia is therefore a borderline outlier (a "true" outlier is usually defined as one that exceeds three standard errors). There is something unique about Alicia. Does she have a special kind of training? Does she write well? know statistics? work with computers? Maybe she just has great "managerial style"? wears "dress-for-success" clothes? plays a great game of golf? Factors such as these have been known to play a role in corporate circles, but here we simply do not know.

Now we check the underlying assumptions of the model. Are the residuals normally distributed? The histogram looks a little like a bell-shaped curve, but it is hard to be sure with such a small sample size. We note Alicia's salary as a near-outlier (beyond two standard errors). Since the chi-square test statistic (1.271) is small, it appears that the errors are normal (the chi-square test statistic ideally would be zero, indicating a perfect normal fit). This means that our confidence intervals should be reliable. We effectively have only four classes ($k = 4$ since the end classes are combined) so that d.f. $= k - 1 - q = 4 - 1 - 2$ (since $m = 2$ parameters were estimated). Mild non-normality is not usually considered a major violation, especially in large samples.

```
......   TEST 1: NORMALITY OF ERRORS   ......

 f(o)      f(e)     z Scale and Histogram
------    ------    +------------------------------
   0       0.0      |
                 -3+|
   0       0.4      |
                 -2+|
   2       2.7      | **
                 -1+|
   8       6.8      | *******
                  0+|
   8       6.8      | *******
                  1+|
   1       2.7      | *
                  2+|
   1       0.4      | *
                  3+|
   0       0.0      |

 Chi-Square =    1.271 with d.f. =   1

 Note: First and last 3 classes collapsed to enlarge f(e).
```

The autocorrelation coefficient (.004) is very close to zero, as would be expected in cross-sectional data. The Durbin-Watson test statistic (1.968) is near 2, also indicating lack of autocorrelation. Since this is not a time series, a nonzero autocorrelation coefficient would not really matter, since it would merely reflect patterns in data entry.

```
...... TEST 2: AUTOCORRELATION OF ERRORS ......

Correlation of e(i) with e(i-1) = 0.004
Computed t for autocorrelation =  0.014 with d.f. =    14
Durbin-Watson =  1.968
```

Heteroskedasticity (nonconstant residual variance) does not appear to be a problem. Bartlett's test statistic (1.214) is small (near the ideal value of zero), so we cannot reject the hypothesis that the errors have a constant variance.

```
...... TEST 3: HETEROSKEDASTICITY OF ERRORS ......

=======================================
Group   Cases    Variance of Residuals
---------------------------------------
  1        10          1241724
  2        10           409204
---------------------------------------

Bartlett's Chi-Square =    1.214
              d.f. =  1
```

However, it does appear that the variance is larger for the first group of residuals. Using the residual variances, *Hartley's F test* can be performed to check for heteroskedasticity:

$$F = \frac{\text{largest residual variance}}{\text{smallest residual variance}} = \frac{1,241,724}{409,204} = 3.03$$

using degrees of freedom for c groups as follows:

Numerator d.f. = c = 2

Denominator d.f. = n/c − 1 = 20/2 − 1 = 10 − 1 = 9.

A large F statistic would indicate unequal residual variances. However, the sample F ratio (3.03) is below the Appendix E critical value (F = 4.26) at the .05 level of significance with d.f. = 2,9 so it appears that the residuals are homoskedastic (constant variance). We summarize our findings in Table 15.3.

The problem with these tests is that they reflect the order of data entry, which would be fine for a time series but not for cross-sectional data such as ours. A better (and simpler) indication of heteroskedasticity may be found by plotting our regression's residuals against each of our predictors separately. For further tests we ask MGRES to save the residuals and predicted Y values in two data files named NEWSAL and SALRES, respectively. Having finished with MGRES, we run PLOT to see the scatter plot of NEWSAL (estimated

Table 15.3   Summary of Residual Test Results at $\alpha = .05$

Possible Residual Violation	Violated?	Critical Value	d.f.	Test Statistic
Non-normality?	No	$\chi^2 = 3.841$	1	$\chi^2 = 1.271$
Autocorrelation?	No*	$t = 2.145$	14	$t = .014$
Heteroskedasticity?	No	$\chi^2 = 3.841$	1	$\chi^2 = 1.214$
		$F = 4.26$	9,9	$F = 3.03$

Autocorrelation test irrelevant except in time-series data.

salary) against SALARY (actual salary) for our twenty employees. The result is shown in Figure 15.7. Given our model's strong predictive power, we are not surprised to find that the fit is nearly perfect (almost forming an ideal 45-degree line, which is the locus of points such that $Y_i = \hat{Y}_i$).

```
OPTION TO SAVE PREDICTED Y

Save predicted SALARY values in a file (y/n)? y
Drive for saving predicted values (A, B, C, or D)? b
File name for predicted values? newsal
File B:NEWSAL successfully saved.


OPTION TO SAVE RESIDUALS

Save residuals in a file (y/n)? y
Drive for saving residuals (A, B, C, or D)? b
File name for residuals? salres
File B:SALRES successfully saved.

Where now:   E=EXPLORE menu    R=Run MGRES again    X=exit ? e
```

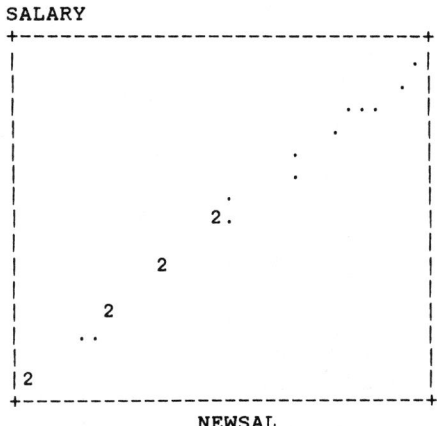

Figure 15.7   Plot of Estimated Against Actual Salary

As shown in Figure 15.8, we also plot the residuals against each predictor variable but find no pattern to suggest heteroskedasticity. Heteroskedasticity would be revealed as a pattern of changing vertical dispersion of the residuals as we move along the lower axis (left to right). If the residuals really have constant variance, the vertical dispersion will be constant, regardless of the predictor value. Slight patterns are observed for SEX EDUC, but they are not conclusive. Note that it is mathematically necessary that the residuals be uncorrelated with each predictor.

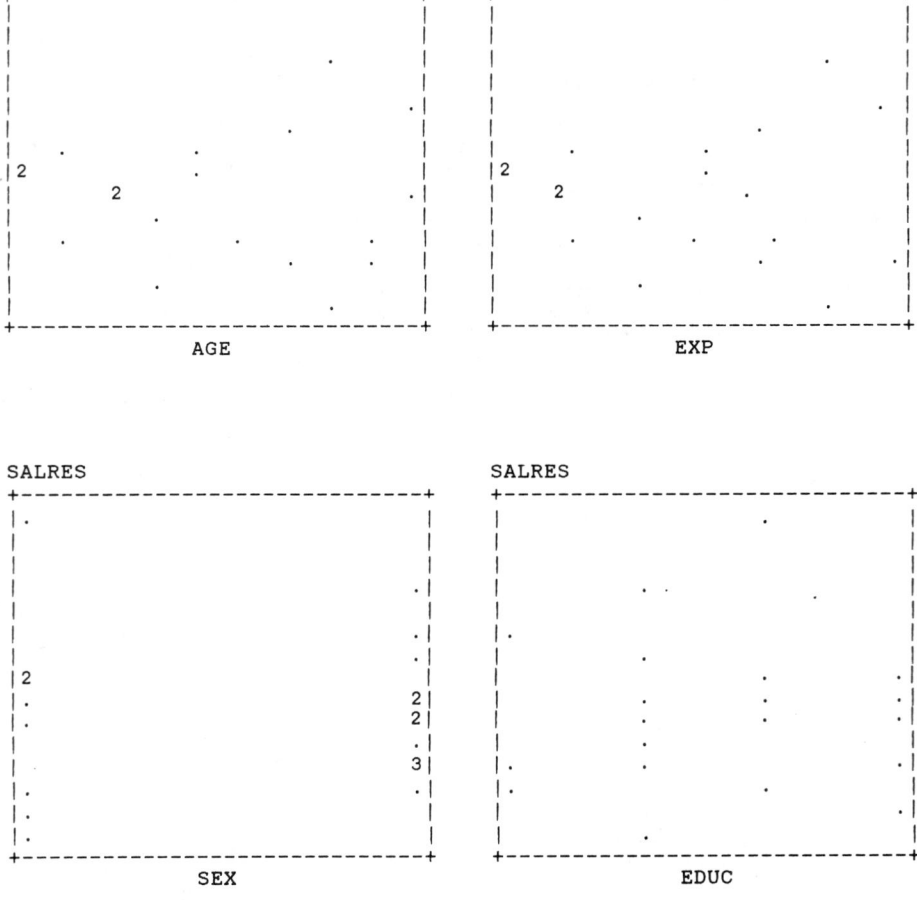

Figure 15.8 Residual Plots Against Individual Predictors

■ Technical Notes for Advanced Learners

The presence of significant pairwise correlation among our predictors (such as AGE and EXP) has already been noted. However, multicollinearity means *any dependence among predictors,* not just bivariate correlation. For example, by running MGRES again we can easily verify that EDUC is highly dependent on EXP and SEX:

$$EDUC = 6.4727 - .1482 \text{ EXP} - 1.935 \text{ SEX} \qquad (r^2 = .7978).$$
$$(t = -5.9) \quad (t = -4.1)$$

This equation says that more experienced workers have less education and that male employees (SEX = 1) have less education than females at Ephemeral Products. Other complex relationships could exist among predictors. This pattern of multicollinearity suggests that we might want to experiment with different combinations of predictors. When we drop a seriously multicollinear predictor from a model, dramatic changes might occur in the estimates of the remaining predictors.

Table 15.4 illustrates a few of the possibilities. For example, when we drop EXP from our original model (Run 2), the coefficients of the other predictors change dramatically (AGE remains significant, while EDUC and SEX show sign reversal and become insignificant). The standard error increases sharply. Our original model estimates therefore appear to be unstable under the impact of discarding EXP as a predictor. This should make us nervous about trusting the original estimates. However, the tradeoff is that the standard error is much lower in the original model.

Table 15.4   Summary of MGRES Experiments, Discarding Various Predictors

Predictor	Run 1 $\hat{\beta}$	t	Run 2 $\hat{\beta}$	t	Run 3 $\hat{\beta}$	t	Run 4 $\hat{\beta}$	t	Run 5 $\hat{\beta}$	t
AGE	1064	20.4	1482	20.8	1518	34.8	1515	37.3	xxxx	xxxx
EXP	822	9.5	xxxx	xxxx	xxxx	xxxx	xxxx	xxxx	xxxx	xxxx
SEX	1296	1.8	−999	−0.6	−298	−0.2	xxxx	xxxx	12096	1.2
EDUC	575	2.2	−383	−0.6	xxxx	xxxx	xxxx	xxxx	xxxx	xxxx
r^2	.9982		.9876		.9872		.9872		.0759	
SE	1049		2682		2635		2565		21794	
Intercept	−2829		−4402		−7549		−7604		49363	

Note: xxxx indicates that a predictor was omitted from a run.

Further experimentation (Runs 3 and 4) shows that once EXP is gone, getting rid of SEX and EDUC does not cause further deterioration in fit (as measured by the standard error) though the t value for AGE becomes more significant as the remaining predictor AGE carries more of the burden of "explaining" variation in Y. As a final experiment (Run 5) we see that regression of SALARY on SEX alone would yield a terrible fit, indicating that SEX alone cannot explain much of the variation in SALARY. Note the dramatic change in the intercept in the last run, as the regression plane is tilted sharply to reflect the regression of SALARY against a single binary predictor.

These experiments are merely to show what can happen and should not be treated as recommended alternatives. If the correct relationship is specified as

$$\text{SALARY} = \beta_0 + \beta_1 \text{ AGE} + \beta_2 \text{ EXP} + \beta_3 \text{ SEX} + \beta_4 \text{ EDUC} + \text{Error},$$

then all of these experiments would involve *model specification error*, so they are open to challenge. We are left with a dilemma. Do we prefer our original model, which has *a priori* reasonable signs and gives good predictions, but whose coefficients appear unstable and may be afflicted by multicollinearity? Or do we prefer possibly misspecified alternative models, which give worse predictions?

A statistician's next step would probably be to estimate separate regressions for the men and women and to compare the coefficients for the two sexes. However, this procedure may raise new problems, too. Or, to cope with multicollinearity, advanced techniques such as *ridge regression* could be tried. That is why further courses in regression are offered to allow you to learn ways of handling such problems. Beginners only need to recognize that these problems exist.

■ Exercises

15.1 From the DATABASE-1 diskette (or another source if you wish), choose any variable Y that you would like to explain. Then choose a reasonable set of predictors (independent variables) that you would expect to affect Y. Use the state database if you want a cross-sectional model, or the time-series database if you want a time-series model. Your model will be of the form:

$$Y = f(X_1, X_2, \ldots, X_m).$$

15.2 Explain carefully the causal relationship that you believe may exist between each predictor and Y. Describe the thinking that led you to choose each predictor. State the *a priori* hypothesis that you hold concerning the sign of each coefficient in your model: Which do you anticipate will be positive and which negative? Why? If you can argue either way about the anticipated sign, explain why the sign is uncertain.

15.3 Discuss whether the actual data do a good job of measuring the underlying cause-and-effect relationships that you propose. If not, what data do you *wish* you had?

15.4 Use the EXPLORE file editor to create files if necessary. If you are using the DATABASE-1 files, just proceed.

15.5 Use LAYOUT to list all the files you are using. If you are using the state database, be sure to include the file STATE (which contains the names of the states) so that you can identify all data points. If you are using the time-series database, be sure to include the file YEAR (which contains the year of each observation). Obtain a printout.

15.6 Use MGRES to estimate the proposed model. If you are using database files, use STATE or YEAR as a label file. Make a printout. Save the residuals and predicted Y values.

15.7 Use PLOT to display a scatter diagram of Y against each predictor. Use PLOT to depict the residuals against each X. Use PLOT to display your estimated Y against the actual Y. Summarize in words what these PLOT runs tell you.

15.8 Informally, look over the regression results, noting the r^2 and t values. Glance at the table of actual and estimated Y. Does it look, at first glance, as if your regression is significant? Are the predictions reasonable? Do *not* do any formal tests—just make a quick scan.

15.9 Study the matrix of correlation coefficients. Is Y highly correlated with the proposed predictors? Which predictors look most promising? Circle all high correlations (using the quick rule or a table of critical values for correlation coefficients).

15.10 Study the off-diagonal elements of the correlation matrix below the top row, looking for signs of multicollinearity among the predictors. Circle the high correlations (using either the quick rule or a table of critical values for correlation coefficients). Which correlations appear to be *significant?* Look at the overall multiple correlation coefficient. Using Klein's Rule, which pairwise predictor correlations are *severe?* What are the implications of your findings for the regression's reliability?

15.11 Choose a level of significance and use the F test to see whether the overall regression is significant. Illustrate your decision rule. Discuss the implications of your decision.

15.12 Choose a level of significance, and test each coefficient in your model to see which ones are significant. Illustrate your decision rule(s), stating your hypotheses clearly. Discuss the implications of these tests.

15.13 Compare the actual sign of each coefficient with the sign you anticipated *a priori*. If there are differences, can you think of reasons?

15.14 Construct a confidence interval for at least one Y value, using the estimated Y from MGRES. Given the size of the standard error and the size of the resulting confidence interval, do you think your model's predictive power is low, medium, or high (using your own definitions)?

15.15 Look over the table of residuals. Circle in red any residuals that are *outliers* (more than three standard errors). In blue ink, circle the residuals that are *near-outliers* (more than two standard errors). In pencil, circle the residuals that are *noteworthy* (more than 1 standard error). Identify the state (or year) in which these residuals occurred. Can you think of reasons why the fit might be poor in these cases?

15.16 Test the residuals for possible non-normality. Choose any level of significance. Discuss the "eyeball" test (comparing observed and expected frequencies visually) as well as the chi-square goodness-of-fit test. Illustrate your decision rule(s) and discuss the implications for your model's reliability.

15.17 If you are using a time series, test the residuals for possible autocorrelation. Choose any level of significance. Use the "eyeball" test for residual patterns in the table of actual and estimated Y values, as well as the Durbin-Watson statistic and t test. Illustrate your decision rule(s) and discuss the implications for your model's reliability.

15.18 Test the residuals for possible heteroskedasticity. Choose any level of significance. Use the "eyeball" test for equal residual variances in the groups as well as the Hartley's F test and Bartlett's chi-square test. Illustrate your decision rule(s). Above all, discuss your plots of the residuals against each X. What is your final conclusion about heteroskedasticity?

15.19 If you encountered a violation of any of the residual assumptions, what (if anything) could be done about it? Was the violation mild or severe?

15.20 If you are ambitious to try some experiments, rerun MGRES, omitting any predictors that were of doubtful importance. Discuss what happens to the overall fit (r^2, standard error, and F). Discuss what happens to the estimated coefficients and t values for the remaining predictors. Do they appear unstable as you try deleting other predictors, or do they remain approximately the same?

15.21 If you think it is necessary, try rerunning MGRES with some additional predictors that you did not include the first time. Discuss their effect on the fit (r^2, standard error, and F) and on the estimated coefficients and t values. Was their inclusion worthwhile?

15.22 Make a table (following the sample format in Table 15.5) to summarize your various MGRES runs. Round off your estimates (to two or three digits) to make the comparisons between columns easier. Discuss the stability of your estimated coefficients and t values as you add or omit predictors. Balancing all factors (simplicity, fit, stability of predictors, logic of cause-and-effect), which model would you recommend?

Sample Format

This example illustrates a typical regression reporting format using a hypothetical model of a college graduate's starting salary. Following MGRES Run 1, the predictor SLOB is discarded, to yield Run 2. This results in only a slight drop in r^2 because the omitted predictor is of marginal significance. Then both SLOB and STATS are omitted, to yield Run 3. This time, r^2 falls noticeably. The t values indicate that the remaining predictors try to pick up the influence of the omitted predictors. The intercept and standard error also shift as various predictors are removed, especially in Run 3, in which a relevant predictor is discarded.

$$\text{MODEL: SALARY} = f(\text{GPA, STATS, SLOB})$$

where

SALARY	=	starting salary (in thousands);
GPA	=	cumulative grade point average;
STATS	=	1 if statistics major, 0 otherwise;
SLOB	=	1 if unkempt, 0 otherwise.

Table 15.5   Estimated Model Coefficients

Predictor	Run 1 $\hat{\beta}$	Run 1 t	Run 2 $\hat{\beta}$	Run 2 t	Run 3 $\hat{\beta}$	Run 3 t
GPA	3.21	6.7	3.41	7.4	4.75	12.3
STATS	1.23	8.5	1.35	9.7	xxxx	xxxx
SLOB	−6.06	−1.0	xxxx	xxxx	xxxx	xxxx
r^2	.458		.412		.258	
Standard Error	3.64		4.71		6.07	
Intercept	7.05		6.55		4.29	

Note: xxxxx indicates a predictor was omitted from the run.

15.23 Write an executive-style synopsis, not longer than two pages, summarizing your main findings. As a supporting exhibit, include your detailed analysis (such as responses to the preceding questions, along with anything else you feel is relevant). Use well-organized, clearly labeled appendices to contain the details of your analysis. As you write, imagine that you will be judged for promotion by your supervisor's supervisor, who will read your report carefully. Be attentive to anything that might cause your report to lose its communication value.

■ References

1. Berenson, Mark L. and Levine, David M., *Basic Business Statistics: Concepts and Applications,* 3rd edition (Englewood Cliffs, N.J.: Prentice-Hall, 1986), pp. 480–481.

2. Neter, John, Wasserman, William, and Kutner, Michael H., *Applied Linear Regression Models* (Homewood, Ill.: Richard D. Irwin, Inc., 1983).

16

MONTE/Central Limit Theorem Demonstration

Introduction

Statistical research relies on mathematical proofs of theorems about the behavior of statistics derived from random samples. But often the real-world implication of these theorems is not terribly clear to the struggling student. Fortunately, it is fairly easy to set up actual experiments to illustrate the validity of the important *Central Limit Theorem*. According to this theorem, the means of random samples tend to follow a normal, bell-shaped curve. Using MONTE, you can subject this theorem to the harshest tests you can imagine.

There is no reason that a computer must be used to illustrate this theorem. You may have seen an exhibit in a science museum in which hundreds of balls or marbles are dropped through a labyrinth of obstacles (similar to a pinball machine). Despite their random paths through this maze, the balls end up at the bottom in a nice bell-shaped heap. Drawing numbers randomly from a hat and calculating the mean produces the same bell-shaped curve. Either of these is a useful way to study the Central Limit Theorem's meaning in real life.

The advantage of using a computer is that it can carry out a random sampling experiment dozens of times in a few seconds and keep track of the results. These results can be summarized visually, so you can see what happened. Perhaps all three methods (dropping marbles, drawing numbers from a hat, running a computer program) are complementary to each other. You might try some of these noncomputer experiments after you have tried MONTE.

MONTE is named after the famous casinos of Monte Carlo, where the laws of chance are routinely explored by nonstatisticians. Such explorations can be costly ways to learn. The generic label *Monte Carlo simulation* refers to any experiment in which random numbers are drawn in order to simulate sampling.

A lot of statistical research in the computer age relies on Monte Carlo simulation because immense calculations and thousands (or millions) of replications are economical—often being much easier than trying to figure things out mathematically. The underlying mathematics for MONTE is not complex, so you can actually check the computer's work if you wish. The arithmetic might be tedious, however.

■ Case Study 16.1: Sampling a Population

The Central Limit Theorem is considered one of the most fundamental statistical laws. It may be verbalized in many elegant ways, but for sample means it may be stated as follows:

> If a population has a mean μ and a standard deviation σ and if random samples of size n are drawn from this population, then the distribution of these sample means approaches a normal distribution with mean μ and standard deviation $\sigma/\sqrt{n}$ as the sample size increases.

Technically, the population must satisfy certain requirements for this theorem to apply, but in practice it is almost impossible to find situations in which it will not work. In an experimental setting we can illustrate this theorem by setting up a population, drawing random samples from it, and seeing how the sample means behave.

This theorem is important because it does *not* say that the population must be normal. Even if your population is "weird," the theorem still applies. *Weird* simply refers to any population that is not bell-shaped. Suppose, for example, that the population is uniform (no peak at all), bimodal (two peaks instead of the normal's single peak), or skewed (one long tail). Figure 16.1 shows these possibilities.

Regardless of the population's shape, the sample means still *tend* to form a normal, bell-shaped histogram. But if the population is quite strange, a histogram of sample means will not be bell-shaped, especially in small samples. In small samples the means do not average out all the irregularities in the parent population, and sample means will replicate, to some extent, the population's shape. To use an extreme case, samples of size n = 1 will look exactly like the population (since you are just pulling a single item from the population). If the population is not too weird, the Central Limit Theorem's predictions will show up clearly, even for modest sample sizes.

With MONTE you can explore various strangely shaped populations of your devising. You can also explore various sample sizes. Start out with small samples, and increase your sample size until you are satisfied that the Central Limit Theorem is beginning to work. The effect of sample size is to increase

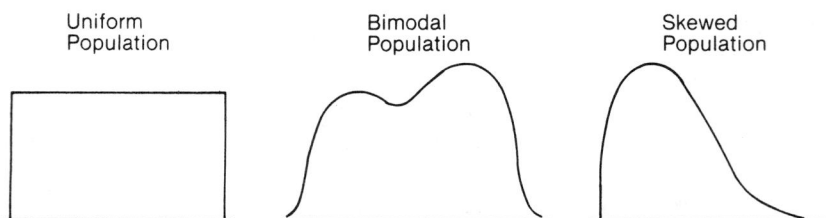

Figure 16.1   Some Non-Normal Populations

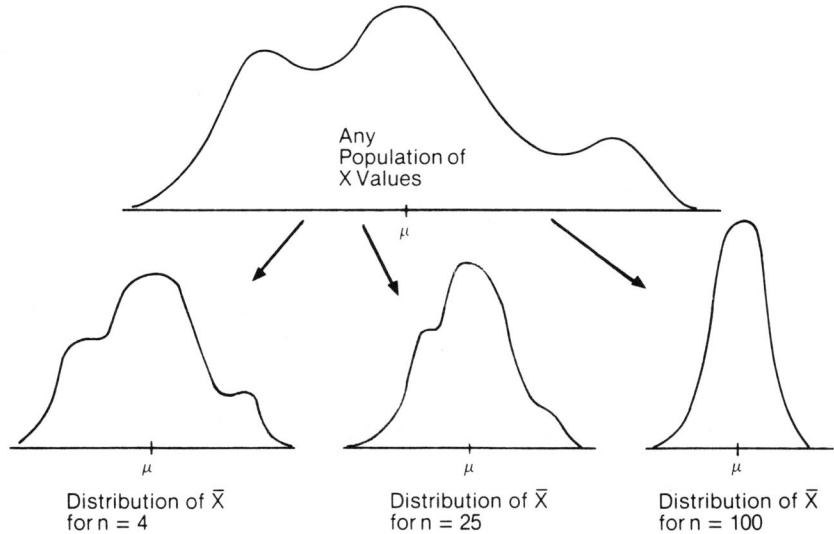

Figure 16.2   Effect of Sample Size on Distribution of X̄

the tendency toward bell shape and simultaneously to collapse the distribution of sample means around the population mean μ, as illustrated in Figure 16.2.

The standard deviation of the sample means $\sigma/\sqrt{n}$ would equal σ when $n = 1$ (same as the population). However, the variance rapidly drops toward zero as n increases. This makes it a little hard to display things visually, because the histogram of means is crowded into a short interval on the X axis. Although the sampling distribution may be bell-shaped, its shape will not be apparent unless we expand the axis scale to spread things out a little. MONTE will do this automatically.

Before proceeding, let's consider how MONTE works. MONTE asks you to specify the general shape of the population histogram. Then, using a random number generator, the computer will choose m random samples, each consisting of n items from your population. This procedure is illustrated in Figure 16.3.

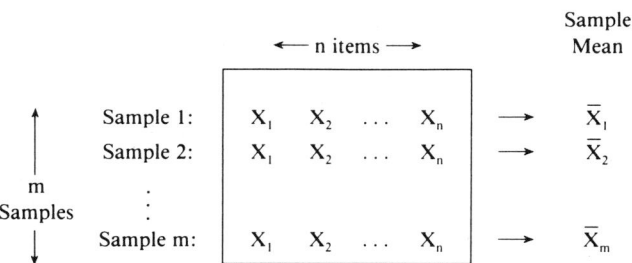

Figure 16.3   MONTE's Sampling Process Illustrated

Each sample will be different. To avoid using additional subscripts, the notation in Figure 16.3 does not show this. You will specify m and n. The mean of each sample is printed, followed by a histogram of the means. In some cases this histogram will be expanded automatically, so you can see more details. MONTE calculates the *mean of the sample means* $\overline{\overline{X}}$ as

$$\overline{\overline{X}} = \frac{\overline{X}_1 \, \overline{X}_2 + \ldots + \overline{X}_m}{m} = \frac{\text{sum of all sample means}}{\text{number of samples drawn}}.$$

According to statistical theory, $\overline{\overline{X}}$ should be close to the true population mean μ. The larger your sample size (n), the less difference you should expect. That is,

$$\overline{\overline{X}} \to \mu \quad \text{as} \quad n \to \infty.$$

Also of interest is the *standard deviation of the sample means:*

$$s_{\overline{X}} = \sqrt{\frac{\Sigma[\overline{X}_j - \overline{\overline{X}}]^2}{m}} = \sqrt{\frac{\text{sum of squared deviations}}{\text{number of samples}}}.$$

As sample increases, $s_{\overline{X}}$ should approach its theoretical limit $\sigma/\sqrt{n}$:

$$s_{\overline{X}} \to \sigma/\sqrt{n} \quad \text{as} \quad n \to \infty.$$

MONTE sets up a population of one hundred values of X. For simplicity, X always lies between 0 and 100. You specify the general shape of the population by giving the frequency for each of ten equal classes. These classes for the values of X may be thought of as cutoff points on a scale:

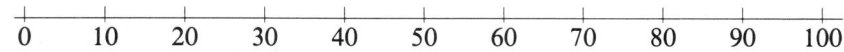

MONTE will ask you how many of the one hundred population X values should be within each of these intervals.

A few examples of such populations will illustrate this procedure. If you wanted a bimodal population histogram, for example, you could specify the distribution shown in Figure 16.4.

If you wanted a skewed population histogram, you could specify something like Figure 16.5.

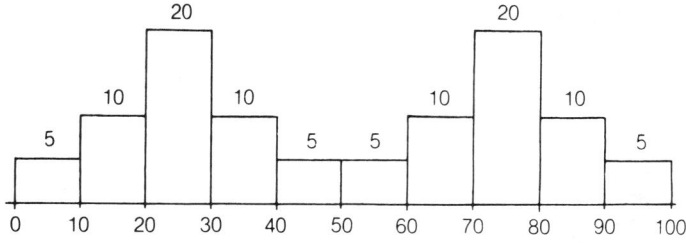

Figure 16.4 Bimodal Population of One Hundred Items

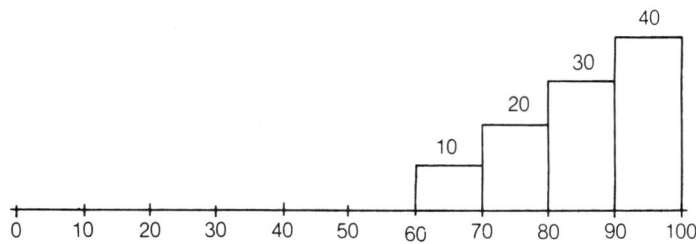

Figure 16.5 Skewed Population of One Hundred Items

Skewed histograms tend to provide the harshest tests of the Central Limit Theorem. Don't be surprised if it takes quite a large sample size to attain a reasonable approach to normality. But you should experiment until you are satisfied. Regardless of the way you set it up, the laws of statistics are invariable.

When you input the frequencies for each of the ten classes, the computer constructs a population vector of one hundred decimal numbers. These numbers are uniformly distributed within each interval. For example, suppose you asked for five population items within the first interval (between 0 and 9.999999). The computer will choose five random decimal numbers that meet this requirement. The X values within each class j (j = 1, 2, . . . , 10) are assigned by using the RND function in BASIC:

$$X(i) = L(j) + 10*RND \qquad \text{where } L(j) = \text{lower limit of class } j.$$

RND is a function yielding a random number between 0 and 1. The five population items in the first interval might be

$$X(1) = 0.514388,$$
$$X(2) = 3.917011,$$
$$X(3) = 4.214386,$$
$$X(4) = 7.371945,$$
$$X(5) = 8.713444.$$

These five numbers will comprise the population's first interval. You cannot see them. All you will see is the histogram, which will show five items between 0 and 10. When a sample is drawn, the computer randomly chooses an integer i between 1 and 100, using

$$i = INT(100*RND + 1)$$

where INT specifies that the number is to be truncated to the next lower integer. (Food for thought: Why do we add 1?) Then we use this index i to pick an item X(i) from our population:

$$X(1), X(2), . . . , X(100).$$

This process is repeated until n items have been chosen. Then the various sample statistics are calculated for our n items. Sampling is *with replacement,* so the same X(i) may be picked more than once.

Let's illustrate this procedure by creating a symmetric, uniform population. We will put the same number of observations in each histogram class. Since there are ten classes and we need one hundred population values, we put ten in each class (see the sample printout for details).

The population histogram appears as expected. The population mean $\mu = 49.23$ is close to 50, as expected. If it seems puzzling that μ isn't exactly 50, remember that frequencies *within* each interval are randomly distributed, and may not be exactly uniform (due to chance). The population standard deviation is $\sigma = 28.69$, which is almost exactly what one would expect for a uniform population (see RAND Case Study 19.2).

To begin with, we choose one hundred samples of four items. The computer prints their means. We know that each mean is the average of four population items. We can see that the means are usually close to the center of the scale (never close to 0 or 100). The histogram shows that indeed the distribution of means looks somewhat like a bell-shaped distribution. In theory the sample means should be near the population mean:

$$\mu = 49.23.$$

In fact, the sample mean of means is

$$\overline{\overline{X}} = 50.78$$

which is pretty close to the expected result. The difference is small enough to be attributed to sample error. Similarly, the standard deviation should be near

$$\sigma/\sqrt{n} = 28.69/\sqrt{4} = 14.35.$$

For the sample,

$$s_{\overline{x}} = 13.46$$

which is pretty close. This agreement tends to give us faith in the Central Limit Theorem, especially since we are using only a small sample size (n = 4).

Now we will increase the sample size to n = 25. The computer prints the mean of our one hundred samples. This time we can see that the means are even nearer the middle of the scale, rarely getting very far away from the population mean of 49.23. The histogram, though, looks bizarre. Because the distribution has collapsed so much, it is hard to see the true shape. However, we can check the convergence toward the theoretical mean:

$$\mu = 49.23$$

which is very close to the sample mean of means:

$$\overline{\overline{X}} = 48.64.$$

For the standard deviation, theory says that

$$\sigma/\sqrt{n} = 28.69/\sqrt{25} = 5.74,$$

which is also very close to the sample result:

$$s_{\bar{x}} = 5.29.$$

The computer offers us an expanded histogram scale because it really is not easy to tell much from the original histogram. From the enlarged-scale histogram we can clearly see the bell shape. The correspondence is still not perfect, but it is good enough to illustrate the point.

You can also explore the Central Limit Theorem using the program SAMPLER (see Chapter 20).

```
                        MONTE

        Everyone agrees that the Central Limit Theorem
(CLT) is true, because it has been proven mathemati-
cally.  But it can also be demonstrated in actual
samples.  This program will let you try as many ex-
periments as you need, until you are satisfied that
sample means DO TEND to form a bell-shaped (normal)
distribution, especially when sample size is large.

        This program draws repeated random samples, with
replacement, from a finite population of 100 elements.
You will specify the general shape of the population
by specifying frequencies for each of 10 equal classes.
X covers a range from 0 to 100.  The computer will
construct an actual population with this shape, find
its mean and standard deviation, and print a histogram
to show the population's shape.  Then, you may draw
as many samples as you wish (up to 100 samples) of any
specified size (not to exceed n=100).  Have fun, as you
test the time-honored laws of statistics!

For random start, please input a 1-digit number? 3


DESIRED NUMBER IN EACH CLASS

Please enter desired frequencies for each of these ten
classes, so the total will be 100 population elements.

From    To (not incl)   Freq
----    -------------   ----
 0           10         ? 10
10           20         ? 10
20           30         ? 10
30           40         ? 10
40           50         ? 10
50           60         ? 10
60           70         ? 10
70           80         ? 10
80           90         ? 10
90          100         ? 10
```

```
                    POPULATION HISTOGRAM
       X VALUE:
       From    To
       ------------------------------------------
         0     10    *********
        10     20    *********
        20     30    *********
        30     40    *********
        40     50    *********
        50     60    *********
        60     70    *********
        70     80    *********
        80     90    *********
        90    100    *********
       ------------------------------------------
```

Population Mean = 49.2297
Population S.D. = 28.69347

Enter the size of each sample (n)? 4
How many samples of size 4 should I draw? 100

```
          MEANS OF 100 SAMPLES (EACH MEAN BASED ON N= 4 ITEMS)

     51.277   45.083   44.288   56.745   36.912   68.696   58.407
     73.137   44.721   41.292   34.287   69.357   18.994   69.949
     72.360   68.330   42.500   59.104   55.422   49.748   48.700
     67.137   64.091   40.018   44.620   42.825   57.125   31.408
     60.586   56.235   47.606   37.054   39.938   44.739   30.975
     29.777   50.843   71.779   26.603   76.237   55.271   51.544
     57.003   39.316   44.234   45.733   37.157   61.871   64.970
     47.880   19.653   50.925   44.728   62.165   59.642   63.629
     58.650   62.918   80.620   34.417   48.801   62.858   42.057
     53.514   60.322   39.557   44.786   36.706   62.509   71.702
     29.996   41.689   28.278   35.701   74.199   45.484   69.120
     49.820   42.130   45.734   33.120   26.961   51.124   46.068
     28.745   61.258   56.519   68.669   54.279   61.430   63.526
     46.096   43.053   65.674   59.912   49.233   60.693   40.936
     46.996   57.282
```

```
       X VALUE:
       From    To      SAMPLE MEAN HISTOGRAM
       ---------------------------------------------------------
         0     10
        10     20    **
        20     30    ******
        30     40    *************
        40     50    *******************************
        50     60    *******************
        60     70    **********************
        70     80    ******
        80     90    *
        90    100
       ---------------------------------------------------------
```

Mean of all sample means drawn = 50.77766
S.D. of all sample means drawn = 13.46454
Number of samples drawn = 100
Size of each sample = 4

Theoretical mean of means = 49.2297
Theoretical s.d. of means = 14.34673

NEXT ROUND OF SAMPLING

Want to draw more samples from this population (y/n)? y
Enter the size of each sample (n)? 25
How many samples of size 25 should I draw? 100

MEANS OF 100 SAMPLES (EACH MEAN BASED ON N= 25 ITEMS)

49.914	48.289	55.730	39.276	47.492	44.944	42.504
53.005	55.941	46.335	38.699	48.232	53.299	60.264
50.701	52.358	43.980	50.590	52.805	60.238	51.685
49.171	46.003	57.900	49.526	46.068	53.427	54.167
50.742	45.275	52.060	51.306	44.342	49.831	40.983
43.793	52.705	50.880	50.278	36.526	45.274	51.158
48.805	48.043	49.255	49.631	49.192	37.226	55.362
40.116	46.934	55.845	52.570	44.162	53.602	60.676
57.251	57.052	42.856	44.418	48.160	45.472	50.051
49.473	40.518	41.765	53.788	43.241	48.833	42.746
42.659	47.559	54.483	42.550	53.686	52.416	42.502
50.343	58.197	49.403	52.205	47.847	56.190	40.839
46.149	46.080	43.524	51.714	44.632	47.290	50.958
45.274	44.500	43.591	48.146	41.623	42.137	49.231
54.834	49.114					

```
X VALUE:
From   To        SAMPLE MEAN HISTOGRAM
---------------------------------------------------------
  0    10
 10    20
 20    30
 30    40     ****
 40    50     ***** 56 *****
 50    60     ***********************************
 60    70     ***
 70    80
 80    90
 90   100
---------------------------------------------------------
Mean of all sample means drawn = 48.64416
S.D. of all sample means drawn = 5.29171
Number of samples drawn = 100
Size of each sample = 25

Theoretical mean of means = 49.2297
Theoretical s.d. of means = 5.738693
```

```
              ADVICE AND COMMENTARY

Your histogram is crowded into only a few classes.
Consequently, it is hard to tell whether or not
the distribution of sample means is really
very bell-shaped.  This is probably caused by
the large sample size. Shall I change the
histogram scale to show more detail (y/n)? y


              ENLARGED HISTOGRAM
X VALUE:
From   To          Histogram
-----------------------------------------------------
 36    39      ***
 39    42      *******
 42    45      ******************
 45    48      *************
 48    51      **************************
 51    54      ****************
 54    57      ********
 57    60      ****
 60    63      ***
-----------------------------------------------------

NEXT ROUND OF SAMPLING

Want to draw more samples from this population (y/n)? n

Where now:  E=EXPLORE menu   R=Run MONTE again   X=exit ? e
```

■ Exercises

General Instructions: Obtain a printout of your main computer results, if possible. Otherwise, keep track of your results from the screen so that you can answer all the questions posed.

16.1 Think of a way to allocate one hundred observations symmetrically in your population histogram. If you wish, you might just start by taking the last digit of the date of your birth, multiplying it by 5, and putting that many observations in both the fifth and sixth classes. Then distribute the remainder any way you wish. For example, suppose you were born on July 15. Then you might assign the frequencies

	like this	or this	or this
	25	5	10
	0	5	10
	0	10	5
	0	5	0
	25	25	25
	25	25	25
	0	5	0
	0	10	5
	0	5	10
	25	5	10
Total:	100	100	100

16.2 Start by taking one hundred samples of size n = 1. The resulting histogram should look like the population. Does it? Compare the mean and standard deviation of the sample means with the population. Discuss your findings fully.

16.3 Now try a small sample size (say, between n = 2 and n = 5). Compare the histogram of sample means visually with the population histogram. Compare the mean and standard deviation of the sample means with the population. Discuss your findings fully.

16.4 Finally, take a large sample (say, n > 15). Compare the histograms, means, and standard deviations as before. Discuss fully.

16.5 The most severe test for the Central Limit Theorem is a badly skewed population. Experiment with your own ideas of weird or unusual populations, and carry out the same steps as those listed in Exercises 16.2 through 16.4.

16.6 If any of your histograms did not look like bell-shaped curves, why not? If any of your sample statistics did not come close to the expected values, based on your population, why not?

16.7 Make a table summarizing each of your experiments, comparing the expected and actual sample statistics. Then discuss the effects of sample size, population shape, and whatever else you think is relevant.

16.8 A student made the comment, "I've disproved the Central Limit Theorem! I came up with a histogram that doesn't look normal at all. Mathematics isn't such a precise subject after all!" Is she right? What would you need to know to assess the possible reasons for her result?

17
PLOT/Scatter Plot for Two Variables

Introduction

The program PLOT will read two data files and print a scatter plot on a grid, along with a few summary statistics such as the correlation coefficient. Many times, a computer-drawn scatter plot of Y versus X will tell you all you need to know. Three general possibilities exist:

1. X and Y show no meaningful correlation (the scatter plot is so widely dispersed that no practical relationship could be argued convincingly).
2. X and Y are somewhat correlated (the scatter plot shows some relationship, though it is muddled or imperfect).
3. X and Y are strongly correlated (the scatter plot shows a near-linear relationship with little disturbance).

Of course, these are not clear-cut categories. But you can usually decide which one comes closest to applying, just by examining the output of PLOT.

If X and Y are correlated, that does not imply cause-and-effect. However, it does suggest that a burden is placed on the statistican to explain why the observed correlation might exist. If X and Y are uncorrelated, yet there is reason to believe that there is a relationship between them, it is incumbent on the statistican to suggest a multivariate model that would be consistent with the lack of observed bivariate correlation.

The relationship could be direct (X and Y covary in the same direction) or inverse (X and Y covary in opposite directions). This can easily be seen from the PLOT output or from the correlation coefficient's sign:

$$\text{Correlation} >> 0 \qquad \text{(direct relationship)}$$
$$\text{Correlation} << 0 \qquad \text{(inverse relationship)}$$

The symbol $>>$ means "significantly greater than"; $<<$ means "significantly less than." One rule of thumb says that a correlation is *probably* significant if the absolute value of the sample correlation coefficient exceeds $2/\sqrt{n}$.

Outliers can be discerned from a PLOT run. Their effect on correlation coefficients (or on any statistical procedure) is adverse. If you believe that outliers may be a problem, you must decide how to deal with them, but detection is the first step. PLOT can also be used to detect *clusters* of data points, which may tell you something important about X and Y.

PLOT's default grid size is chosen to fit on one screen. You can vary this grid, but if you increase it vertically, it may not look right. When a plot is printed, its proportions may not look the same as what you see on the screen, so a bit of experimentation may be needed.

■ Case Study 17.1: Bond Prices versus Coupon Rates

We take a sample of bonds listed on the New York Stock Exchange (NYSE) by choosing every twentieth bond. If the bond is convertible or the company is in receivership or otherwise incomparable, the next bond is chosen. This procedure results in a sample of 80 bonds. For each bond we record the coupon rate (or original interest rate), year of maturity, and closing price on Friday, July 18, 1986, as shown in Table 17.1. The price is expressed as a percent of face value (par). A bond listed at 100 is selling at face value, those below 100 are said to be selling at a discount, and those above 100 are selling at a premium.

Next we make a scatter plot to view the relationship between the prices of the bonds and their coupon rates. Theory says that the higher the coupon rate (original interest rate), the higher the price should be, other things being equal. In real life, of course, other things (such as risk and maturity) rarely are equal, but a scatter plot should at least give some indication of the pattern.

We use PLOT to produce the printout shown at the end of this chapter. Note that we have shortened the horizontal axis to fit better on the page and have elongated the vertical axis slightly, instead of using the default values.

As anticipated, the plot shows a strong positive correlation (.6421) with two rather distinct outliers (Zapt at 42 and Petrie at 133). The price of Zapt (42) is far below par, despite its relatively high coupon rate (10.25 percent). The price of Petrie (133) is far above par, despite its near-average coupon rate (8.00 percent). We should look closely at these two companies to see what might explain their unusual positions outside the main cluster. Otherwise, the somewhat curvilinear pattern is fairly clear. There are a few other near-outliers, which might also be worth examining.

The correlation coefficient (.6421) is highly significant, since it exceeds our rule of thumb value:

$$2/\sqrt{n} = 2/\sqrt{80} = .2236.$$

But remember that a bivariate correlation is only part of the story. Other factors really should be taken into account if we wish to explain variation in bond prices. Probably, a multiple regression model would be needed to study the

Table 17.1   Yields and Prices of 80 Randomly Selected NYSE Bonds

Firm	Coupon	Mature	Price	Firm	Coupon	Mature	Price
AlaP	8.75	2007	94.00	KaufB	12.25	1999	103.00
AlldC	5.20	1991	86.75	LearS	10.00	2004	101.13
Alcoa	9.00	1995	100.13	LouGs	9.25	2000	101.00
ACyan	7.38	2001	84.00	MarCor	6.50	1988	88.50
Ames	10.00	1995	100.13	MerL	11.68	1987	104.25
AshO	6.15	1992	88.88	Mobil	8.50	2001	94.13
AvcoC	7.50	1993	92.50	MtSTl	9.68	2015	101.75
Banka	7.88	2003	83.25	NtMed	12.75	2000	106.88
BellCn	9.00	2008	98.00	Navstr	14.50	2001	114.75
Blair	13.68	1998	101.88	NiMp	8.35	2007	86.68
BurNo	8.60	1999	99.88	NwnBl	7.50	2005	88.13
CaroT	7.75	2001	89.50	OhBlT	9.00	2018	97.88
ChNY	8.40	1999	97.00	Ownlll	7.68	2001	87.00
ChvrnC	11.00	1990	107.25	PGE	8.00	2003	90.00
ChryF	9.80	1988	103.50	PSwAir	6.00	1987	95.00
CitSv	6.13	1997	72.00	PAA	11.25	1986	100.00
ColuG	7.50	1997	88.25	Petrie	8.00	2010	133.00
CmwE	8.00	2003	88.50	PhilEl	11.00	2000	101.00
ConEd	4.75	1990	90.13	PhillP	12.88	1992	103.00
CnPw	5.88	1996	73.25	PotEl	11.88	1989	106.50
CrnPd	5.75	1992	88.25	PSNH	15.75	1988	106.68
CritAc	11.25	2015	101.25	RJR	7.38	2001	88.00
DetEd	6.00	1996	81.50	Revlin	11.75	1995	100.25
Dow	8.68	2008	93.00	SanD	10.00	2006	102.25
DugL	5.00	2010	59.75	SearA	8.38	1986	100.00
Ens	9.75	1995	99.13	Socny	4.25	1993	80.00
FinCpA	6.00	1988	90.00	Spiegl	5.00	1987	97.50
FroC	8.50	1991	101.50	Subne	5.50	1992	81.00
Fuqua	7.00	1988	97.00	Tenco	14.50	2006	111.50
GEICr	11.75	2005	108.13	TexC	13.68	1994	109.00
GMA	8.75	2000	99.13	TxOG	16.68	1991	110.75
GMA	11.75	1989	105.00	TWA	4.00	1992	59.50
Gene	15.25	1994	99.00	TCFox	13.25	2000	104.68
GaPw	16.13	2011	112.50	UnEl	10.50	2005	104.00
GtNor	2.68	2010	40.00	USBO	7.75	2002	85.00
GlfRes	12.50	2004	99.75	ValerNG	16.75	2002	112.00
Honey	9.38	2009	100.00	WellF	8.60	2002	97.00
ITTF	8.10	1992	97.00	WUTI	9.25	1997	75.00
IngR	8.05	2004	88.00	Woolw	7.38	1996	95.00
IntTT	8.90	1995	100.68	Zapt	10.25	1997	42.00

Source: New York Times, July 20, 1986, pp. 12F–24F. Data are for educational use only and are not intended as a guide for investment decisions.

causes of price variation. Yet it is clear that the coupon rate is likely to be an important part of the explanation.

Lack of significant bivariate correlation does not signify the absence of a relationship, however. It is quite possible for a predictor to be significant in a multivariate model, even though it is uncorrelated in a bivariate model. It is even possible for a bivariate correlation to be negative and significant when the same predictor exhibits a significant *positive* relationship in a multivariate model. For example, the plot of SALARY against EDUC in Chapter 16 clearly shows an inverse relationship, yet the t value for EDUC in the model

$$SALARY = f(AGE, EXP, SEX, EDUC)$$

is both positive and significant. The moral is that PLOT results can be misleading if they are interpreted naively.

```
                              PLOT

          A scatter plot will give you a general idea of
     the relationship between any two variables X and Y.
     Program expects you to have two files already saved.
     The scaling and labeling of your graph is automatic.
     Origin of your graph is (Xmin, Ymin).  The grid size
     for the plot is 50 across by 15 down, which fits well
     on the screen.  However, you can choose any size up to
     65 across by 18 down.  Maximum is N = 1000 cases.


     SET UP YOUR PLOT CHARACTERISTICS

     Shall I use graphics characters (y/n)? n
     Is standard 50 x 15 grid size O.K. (y/n)? n
     Input new grid size (maximum 65 across, 18 down).
     Size across? 40
     Size down? 17

     Which file goes on the X-axis ? coupon
     Which file goes on the Y-axis ? prices


     GENERAL SUMMARY

     Number of observations =   80

     Mean of COUPON =   9.259625
     Mean of PRICES =   94.55049

     Correlation coefficient = 0.6421

     For COUPON: each space =  .35175 units
     For PRICES: each space =  5.470588  units

     COUPON goes from  2.68  to  16.75
     PRICES goes from  40  to  133
     Origin of graph is ( 2.68 , 40  )
```

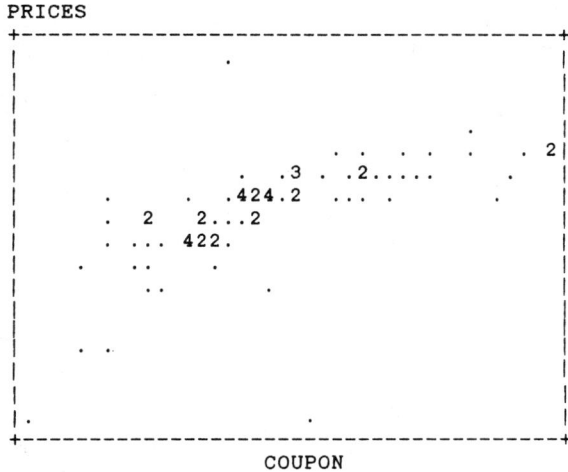

```
Key: Each dot shows one point.  Digits 2,3,...,9 show multiple
     dots at same grid point. Some other symbols MAY be needed:

        a = 10-19 dots              c = 50-99 dots
        b = 20-49 dots              d = 100 or more dots


     PRICES
     +--------------------------------------+
     |                     .                |
     |                                      |
     |                                      |
     |                        . .  .    .  .| 2 |
     |               .   .2.....       .  2 |
     |          . .3 . .2.....       .      |
     |      .      . .424.2  ... .          |
     |    .  2   2...2                      |
     |    . ... 422.                        |
     |      .    ..    .                     |
     |           ..         .                |
     |                                      |
     |                                      |
     |     . .  ..                          |
     |                                      |
     |                                      |
     | .                  .                 |
     +--------------------------------------+
                     COUPON
```

Where now: E=EXPLORE menu   R=Run PLOT again   X=exit ? e

Exercises

17.1 From the DATABASE-1 diskette, choose any two state data files that you think may be related. Use Appendix H to help you choose your files. Use PLOT to obtain a scatter diagram.

17.2 In your own words, write a brief description of the scatter plot's appearance. Discuss the direction of the relationship (if any), outliers, and anything else that is evident to you. If there are outliers, try to identify them and suggest an explanation.

17.3 Try to explain what the scatter plot shows. If there appears to be no relationship, could there nonetheless be a multivariate cause-and-effect model that would allow X to affect Y? If so, what else might the model include?

17.4 Test the correlation coefficient for significance, using the quick rule that a correlation probably is significant if its absolute magnitude exceeds $2/\sqrt{n}$. If there appears to be a significant correlation, could it actually be spurious or misleading? Why?

17.5 Repeat Exercises 17.1 through 17.4 using two time-series variables (from the DATABASE-1 diskette) that you think may be related.

17.6 "A picture is worth a thousand words." Do you agree? Be specific, using your results from PLOT. If you were given only the numerical value of the correlation coefficient, without a scatter plot, what might you overlook?

17.7 A user of PLOT said, "The computer's scatter plot does not label the axes, so the graph is not useful." To what extent is her statement true? Why might a busy data analyst not care about labeling the axes?

18
POIS/Poisson
Probability Model

Introduction

Imagine that you are inspecting a pipeline, looking for leaks. Or imagine that you are attending a serious play, and someone in the audience keeps coughing at inappropriate moments. Or imagine that you get cut off in the middle of a telephone conversation, owing to technical problems on the line. Or imagine watching a baseball game, hoping to see a home run. What do these seemingly unrelated events have in common?

All are possible examples of *Poisson processes,* in which the variable of interest is the number of times something happens *over an interval of time or space.* This model is widely applicable. To visualize a Poisson process, consider Figure 18.1. The line represents a continuum of time (or space), and each asterisk (*) represents an event occurring at a certain point.

Poisson events (shown as asterisks) are randomly positioned in time or space. Let X be the number of events that happen to occur within a randomly chosen unit of time or space. You can see that X depends on where we choose our interval. In Figure 18.1, X = 2 because two asterisks lie within our unit of time. But if we were to slide our unit of time to the left slightly, we would raise X to 4 or 5 or 6. Therefore X is a random variable.

It is important that Poisson events be independent of one another. That is, the probability of one event's occurrence should not be affected by any other. Does one pipeline leak occur near the next one? Does one cough lead to another? Does one telephone disconnection cause another? Do home runs come in clusters? Although this requirement may be difficult to meet precisely, the Poisson model has been found to give a rather good description of many real-life situations.

The Poisson distribution has one parameter μ that represents the mean number of events per unit of time or space. The Poisson probability density function (p.d.f.) depends solely on μ:

$$P(X) = \frac{\mu^X}{X!} e^{-\mu} \qquad \text{where } e = 2.71828.$$

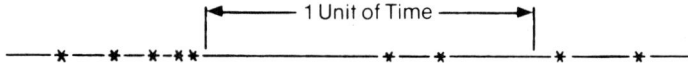

Figure 18.1   Schematic of a Poisson Process

POIS uses a logarithmic transformation to ensure high accuracy in evaluating Poisson probabilities. Consequently, you can experiment over a wide range of situations, perhaps comparing Poisson probabilities with the binomial or normal models.

A Poisson distribution's shape may be evaluated by using these formulas:

$$\text{Skewness} = 1/\sqrt{\mu};$$
$$\text{Kurtosis} = 3 + 1/\mu.$$

POIS calculates these shape parameters and prints them for you. As the mean increases, the skewness parameter approaches 0 (a skewness coefficient of 0 indicates symmetry), and the kurtosis coefficient approaches 3 (a kurtosis coefficient of 3 indicates a bell-shaped curve). This illustrates the fact that a Poisson p.d.f. approaches the normal p.d.f. as the mean increases.

■ Case Study 18.1: Distribution of Home Runs

Suppose that home runs occur for a particular baseball team at an average rate of 1.2 home runs per game. The Poisson p.d.f. is not hard to calculate, but it is tedious, so many statisticians rely on tables. Manual calculations can be awkward (or impossible) when μ or X is large. Using POIS, it is easy to find exact or cumulative Poisson probabilities, as the sample printout shows for this baseball example.

The Poisson probability density function (p.d.f.) for home runs is illustrated in Figure 18.2. The probabilities in Figure 18.2 came from the sample printout on page 200.

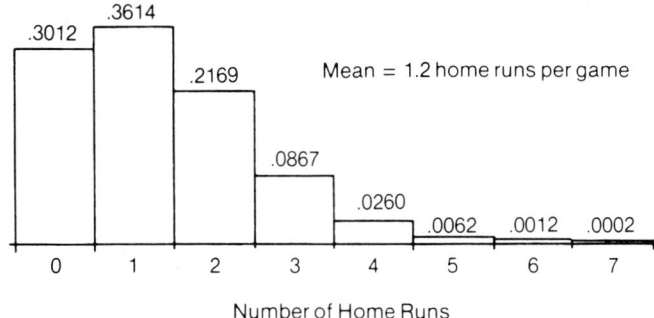

Figure 18.2   Poisson P.D.F. for μ = 1.2

```
                           POIS

This program calculates these Poisson probabilities:

     1. Exact Poisson probabilities
     2. Right-tail cumulative Poisson probabilities
     3. Left-tail cumulative Poisson probabilities

Small probabilities are suppressed to shorten the table.
You will be asked to enter the single Poisson parameter:

        Mu = expected (mean) number of events
              per unit of time or space·

What is the mean (Mu)? 1.2

For this model: Skewness   =     0.913
                Kurtosis   =     3.833


           TABLE OF POISSON PROBABILITIES

   Mu =  1.2  =  mean events per unit of time/space
   r  =  number of events in a particular interval

            Exact           Cumulative      Cumulative
      r     P(X=r)          P(X>=r)         P(X<r)
     ----   ------          -------         ------
      0     0.3012          1.0000          0.0000
      1     0.3614          0.6988          0.3012
      2     0.2169          0.3374          0.6626
      3     0.0867          0.1205          0.8795
      4     0.0260          0.0338          0.9662
      5     0.0062          0.0077          0.9923
      6     0.0012          0.0015          0.9985
      7     0.0002          0.0003          0.9997

Probabilities less than .0001 have been omitted.

Where now:  E=EXPLORE menu   R=Run POIS again   X=exit ? e
```

We have no way to set an a *priori* upper limit on X. But notice that the probabilities fall off rapidly, reaching almost zero beyond 6 or 7. Also, the chance of no home runs (X = 0) is substantial. This accords with our experience (for most teams, at least). Of course, each team would have a different mean (μ). The calculation of the probability of 2 home runs is illustrated below:

$$P(X) = \frac{\mu^X}{X!}\, e^{-\mu}$$

$$P(2) = \frac{1.2^2}{2!}\,(2.71828)^{-1.2} = .2169.$$

In this example the distribution's shape is given by

$$\text{Skewness} = 1/\sqrt{\mu} = 1/\sqrt{1.2} = .913$$

$$\text{Kurtosis} = 3 + 1/\mu = 3 + 1/1.2 = 3.833.$$

The p.d.f. is therefore skewed to the right and is leptokurtic (more peaked than a normal, bell-shaped curve). Figure 18.2 supports this conclusion.

In reality, home runs may not be Poisson events. If home runs occur in clusters, the mean may not be constant from inning to inning. But the Poisson model is certainly a candidate to describe events such as home runs, hockey goals, and similar sporting phenomena.

Case Study 18.2: Poisson versus Binomial Models

The binomial and Poisson models are alike in many ways. Both are models of discrete variables, and X is the number of occurrences in both models. Their p.d.f.'s often look alike. But in the binomial model, X has a clear upper limit (n), while in the Poisson model X has no obvious upper limit.

Of course, we cannot really have an infinite number of home runs (or coughs or leaks or phone disconnections). But in real-life situations the Poisson p.d.f. rapidly tapers off to the point at which probabilities become zero, so for many practical purposes the lack of an upper limit is unimportant. Table 18.1 has been constructed to compare these two models.

What are the rules for using these models? You must look at the situation to see which one fits more closely. Some statisticians call the Poisson a model of rare events, system failures, or defects because it is commonly applied to equipment malfunctions, accidents, or errors. The Poisson is also called the arrival model because it characterizes the arrival of telephone calls at switchboards, patients at emergency rooms, and customers at fast-food restaurants.

A whole branch of management science is devoted to *queueing theory,* which analyzes waiting lines of customers in service facilities such as banks, post offices, and sporting events. In most of these applications, Poisson arrivals are assumed (which is usually a realistic assumption).

You can always approximate a binomial using the Poisson, by setting $\mu = np$. For example, if $n = 1000$ and $p = .001$, it would be difficult to calculate the binomial. But you could set $\mu = 1$ and use the Poisson, since $np = 1000(.001) = 1$. The resulting Poisson probability should be a good approximation to the binomial.

Table 18.1   Comparison of Binomial and Poisson Models

Binomial	Poisson
Two parameters (n and p)	One parameter (μ)
$X = 0, 1, 2, \ldots, n$	$X = 0, 1, 2, \ldots$
$E(X) = np$	$E(X) = \mu$
$V(X) = np(1 - p)$	$V(X) = \mu$

Specifically, the probability of X = 0 successes using the binomial is

$$P(0) = .3677 \quad \text{(binomial with n = 1000, p = .001)},$$

which is obtained from the BINOM example (see Case Study 7.1).

The Poisson approximation we obtain from POIS (see sample printout) is

$$P(0) = .3679 \quad \text{(Poisson with } \mu = 1\text{)},$$

so the approximation is extremely close. As a general rule, the Poisson approximation is best when n is large and p is small, as in this example.

In this example the distribution's shape is given by

$$\text{Skewness} = 1/\sqrt{\mu} = 1/\sqrt{1} = 1.000$$

$$\text{Kurtosis} = 3 + 1/\mu = 3 + 1/1 = 4.000.$$

The p.d.f. is therefore skewed to the right and is leptokurtic (more peaked than a normal, bell-shaped curve). You could sketch the p.d.f. using the P(X = r) column from the POIS printout to verify the shape.

```
                         POIS

     This program calculates these Poisson probabilities:

          1. Exact Poisson probabilities
          2. Right-tail cumulative Poisson probabilities
          3. Left-tail cumulative Poisson probabilities

     Small probabilities are suppressed to shorten the table.
     You will be asked to enter the single Poisson parameter:

          Mu = expected (mean) number of events
               per unit of time or spae

     What is the mean (Mu)? 1

     For this model: Skewness  =     1.000
                     Kurtosis  =     4.000


               TABLE OF POISSON PROBABILITIES

       Mu =  1  =  mean events per unit of time/space
       r  =  number of events in a particular interval

                 Exact          Cumulative      Cumulative
           r     P(X=r)         P(X>=r)          P(X<r)
          ----   ------         -------          ------
           0     0.3679         1.0000           0.0000
           1     0.3679         0.6321           0.3679
           2     0.1839         0.2642           0.7358
           3     0.0613         0.0803           0.9197
           4     0.0153         0.0190           0.9810
           5     0.0031         0.0037           0.9963
           6     0.0005         0.0006           0.9994
           7     0.0001         0.0001           0.9999

     Probabilities less than .0001 have been omitted.

     Where now:  E=EXPLORE menu   R=Run POIS again   X=exit ? e
```

Exercises

18.1 Try running POIS using the following values of the mean:

    a. $\mu = .005$
    b. $\mu = .05$
    c. $\mu = .5$
    d. $\mu = 5$
    e. $\mu = 50$

18.2 Discuss the shape of the p.d.f. in these POIS runs. Sketch the p.d.f.'s roughly, showing where the mean lies. Would you say that any of them is acceptably like a bell-shaped curve? Explain fully.

18.3 Find the standard deviation σ of each of these Poisson models. Construct a two-sigma interval of the form $\mu \pm 2\sigma$, and tell what its upper and lower limits are.

18.4 Use BINOM to find the binomial probabilities for each of the following cases:

    a. n = 1000,   p = .0005
    b. n = 100,   p = .005
    c. n = 50,   p = .01
    d. n = 10,   p = .05

What is the mean for each of these binomials? Compare the results of each BINOM run with the POIS probabilities you obtained in Exercise 18.1c. Are they nearly the same? Should they be? In which cases are the differences most noticeable? What generalization can you make about the effects of n and p on the validity of the Poisson approximation to the binomial?

18.5 Use BINOM to find the binomial probabilities for each of the following cases:

    a. n = 1000,   p = .005
    b. n = 100,   p = .05.
    c. n = 50,   p = .10
    d. n = 10,   p = .50

What is the mean for each of these binomials? Compare the results of each BINOM run with the POIS probabilities you obtained in Exercise 18.1d. Are they nearly the same? Should they be? In which cases are the differences most noticeable? What generalization can you make about the effects of n and p on the validity of the Poisson approximation to the binomial?

18.6 Experiment with weird Poisson p.d.f.'s. Then describe what you have learned about the program POIS as a result of your experiments.

■ Reference

Hastings, N. A. J. and Peacock, J. B., *Statistical Distributions* (New York: John Wiley and Sons, 1975).

19 RAND/Random Number Generator

Introduction

The program RAND will generate three types of random numbers:

1. Normally distributed random numbers with mean μ and standard deviation σ (you choose μ and σ);
2. Uniformly distributed random integers lying between specified limits A and B (you choose A and B);
3. Poisson distributed random numbers with mean μ (you choose μ).

Although statisticians study and use random numbers of other types, these are probably the most common. All samples are drawn with replacement, which means that the same random number could occur more than once. RAND also offers the option of saving the resulting random numbers in a data file.

If you save the random numbers in a data file, you can study the sample (using the programs BOXPLOT or ANALYZ) to see whether the sample really follows the desired model (uniform or normal or Poisson) by looking at the sample histogram, mean, standard deviation, skewness, kurtosis, and so on.

To facilitate the study of skewness and kurtosis, RAND prints the Pearsonian theoretical shape coefficients for the population being sampled:

	Skewness	Kurtosis
1. Normal	0	3
2. Uniform	0	1.8
3. Poisson	$1/\sqrt{\mu}$	$3 + 1/\mu$

It is an interesting exercise to see how close a particular sample comes to the expected shape. As sample size increases, the resemblance will improve regardless of the model. It is also instructive to note that as μ increases, the Poisson shape approaches the normal (although any Poisson distribution is somewhat right-skewed and leptokurtic).

You can also do a formal goodness-to-fit test (using the program GOODFT) to test the saved data file, to see whether the data really fit the desired model (uniform, normal, or Poisson). The data files generated by RAND are ideally suited for GOODFT because the models allowed are the same.

Running ANALYZ, BOXPLOT, and GOODFT with RAND-generated data files can be a useful learning exercise because you know *in advance* what the sample is supposed to look like. Naturally, the sample will never have *exactly* the intended characteristics because of the laws of chance—sample variation does occur.

One option merits special mention because it must often be performed by the statistician: using RAND to select a random sample from a population. Assuming that you have a sequentially numbered population of N items, how can you choose n items at random? Just use RAND's first option, letting A = 1 and B = N. The sample is drawn with replacement, so the same random number could occur more than once. If you need to choose a random sample without replacement, you can draw a slightly larger sample of random numbers than you actually need, then discard any duplicate random numbers. The odds against duplicates depend on the situation (that is, it depends on N and n). This option is illustrated in Case Study 19.1.

Depending on your computer, you may not be able to duplicate the exact sequences of random numbers shown in the case studies. The format, however, will be exactly the same.

■ Case Study 19.1: Sampling Hospital Records

Carver Memorial Hospital wants to know whether proper health insurance reimbursement guidelines were followed in 2452 outpatient psychiatric visits during the past year. To verify a reimbursement, a patient folder must be pulled and audited by a trained accountant. Often a telephone call to the insurance provider is required. Examining all of the outpatient psychiatric visits would be prohibitively expensive. Instead, the hospital can select a random sample of outpatient visits. We can estimate the mean reimbursement per visit from this sample, using the rules of statistical inference to construct a confidence interval for the true mean reimbursement for the population.

Since there is no point in auditing the same patient visit more than once, we should sample without replacement. Assume that the required sample size is determined to be twenty-five outpatient visits. If we draw twenty-five random numbers ranging from 1 to 2452, it is intuitively clear that many duplicates are unlikely, yet it would not be surprising to see one or two. Just to be on the safe side, we will draw thirty random numbers (setting A = 1 and B = 2452) and then discard any duplicates.

The printout on page 208 shows the resulting sample. It happens that there are no duplicates in the first twenty-five random numbers, so the five extra numbers are not needed. Now, assuming that each outpatient visit is numbered from 1 to 2452, the selected patients' files can be pulled, and the reimbursement for each of these twenty-five randomly chosen visits can be recorded:

Sample item 1: Reimbursement for patient # 775

Sample item 2: Reimbursement for patient #2434

.

.

.

Sample item 25: Reimbursement for patient # 900

While the resulting sample mean reimbursement will not be exactly equal to the true population mean, it will be an unbiased estimate. Using standard statistical procedures, upper and lower limits can be established around the sample mean, to achieve whatever degree of confidence is desired. Sample size can be varied scientifically as required.

This type of procedure can save a lot of money in auditing, operations management, and production because it eliminates the need for 100 percent inspection. Instead, statistical control charts can be set up to monitor compliance with intended goals. If problems are detected, corrective steps can be taken immediately. Organizations with ongoing sample inspection programs can devote more resources to correcting their problems, while those that attempt 100 percent inspection may find that the warning comes too late (and at greater cost). The Japanese have utilized this principle to great advantage in the last twenty-five years, and many U.S. manufacturing firms are now attempting to implement programs of statistical quality control.

```
                          RAND

          This program will generate up to 1000 random numbers
     and save them in a file, if you wish.  The options are:

          1. Normal
          2. Uniform
          3. Poisson

     Which option do you want (1, 2, or 3)? 1
     How many random numbers do you want? 30
     For random start, enter any 2-digit number: 33


     PROCEDURE TO MAKE  30 UNIFORM RANDOM NUMBERS

     This program will create 30 equiprobable random integers
     ranging from A to B.  You must supply the values A and B.
     The end points A and B will be included in the set of
     possible random integers.  The resulting random integers
     may be used to select a simple random sample from a
     population, or may be studied (if saved) using another
     program such as ANALYZ or BOXPLOT or GOODFT.

     What is the lower limit A? 1
     What is the upper limit B? 2452
```

```
Note: For a Uniform Model:

Theoretical skewness = 0.000 (Pearson coefficient)
Theoretical kurtosis = 1.800 (Pearson coefficient)

                    30 RANDOM INTEGERS FROM 1 TO 2452

    775  2434   708  2411  1104  1725   168    84  1573   602
    842  1861   180   938  2167  1308  1993   556  1018  2438
   1257   319  2320  2282   900  2020   842  1353   767  1986


OPTION TO SAVE THE DATA

Do you want to save these random numbers (y/n)? n

Where now:  E=EXPLORE menu   R=Run RAND again   X=exit ? e
```

■ Case Study 19.2: Uniform Random Numbers

To see whether the program RAND is truly generating uniformly distributed random numbers, we ask for a sample of fifty random integers from 1 to 10 (as shown on the sample printout on page 210. Our sample histogram would be expected to resemble Figure 19.1.

In fact, the hand-drawn sample histogram shown in Figure 19.2 has noticeably too many 4s and 5s, and too few 7s and 8s.

Of course, we do not expect our sample to be *exactly* uniform. A goodness-of-fit test is needed, using GOODFT, to decide whether our RAND-generated file UNIF33 differs *significantly* from a uniform distribution. Skipping the details of the GOODFT test (not shown here), we will simply say that this observed difference is well within the range of chance and is not a significant departure from uniformity.

Let's examine the sample further. By running ANALYZ on the file UNIF33 we calculate a few summary statistics (ANALYZ printout is not shown). For our fifty random numbers the sample mean and median are

$$Mean = 5.3$$

$$Median = 5.$$

```
    *    *    *    *    *    *    *    *    *    *
    *    *    *    *    *    *    *    *    *    *
    *    *    *    *    *    *    *    *    *    *
    *    *    *    *    *    *    *    *    *    *
    *    *    *    *    *    *    *    *    *    *
   _____
    1    2    3    4    5    6    7    8    9    10
```

Figure 19.1   Ideal Uniform Sample of Size n = 50

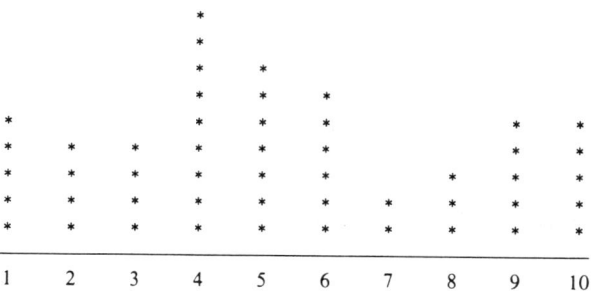

Figure 19.2   Actual Sample of Size n = 50

If the sample were exactly really uniform, the mean should be

$$\mu = \frac{A + B}{2} = \frac{1 + 10}{2} = 5.5.$$

Since a uniform distribution is symmetric, the median should be equal to the mean. Both the mean and median are very close to their expected values. For a true uniform discrete distribution the standard deviation should be equal to

$$\sigma = \sqrt{\frac{[(B - A) + 1]^2 - 1}{12}}.$$

Since A = 1 and B = 10 for our experiment, the true standard deviation should be

$$\sigma = \sqrt{\frac{[(10 - 1) + 1]^2 - 1}{12}} = 2.87.$$

For our sample of fifty random integers saved in file UNIF33, ANALYZ reveals that the sample standard deviation is 2.79. We would have to say that the sample has approximately the expected standard deviation. This tends to confirm that the random number process is working correctly.

The first quartile should cut off the bottom 25 percent of the random numbers (should be about 3), and the third quartile should cut off the top 25 percent (should be about 8). The second quartile is the same as the median (should be about 5.5, the same as the mean). We obtain the actual sample quartile points by running ANALYZ on the data file UNIF33:

First quartile = 3

Second quartile = 5

Third quartile = 8.

Given the discrete nature of the integer data, the sample quartile points come as close as we could expect.

For a true uniform distribution, statistical theory says that the data should be *symmetric* and *platykurtic*. Specifically, the skewness and kurtosis coefficients should be

$$\text{Skewness} = 0$$

$$\text{Kurtosis} = 1.8.$$

Actually, for our file UNIF33, ANALYZ calculates that

$$\text{Skewness} = 0.221$$

$$\text{Kurtosis} = 2.008.$$

Therefore our sample is slightly right-skewed (skewness > 0) and more peaked than a uniform distribution (kurtosis > 1.8). However, these differences are slight and would not cause us to doubt that the numbers are drawn from a uniform population.

```
                          RAND

        This program will generate up to 1000 random numbers
    and save them in a file, if you wish.  The options are:

        1. Normal
        2. Uniform
        3. Poisson

    Which option do you want (1, 2, or 3)? 2
    How many random numbers do you want? 50
    For random start, enter any 2-digit number: 33


    PROCEDURE TO MAKE  50 UNIFORM RANDOM NUMBERS

    This program will create 50 equiprobable random integers
    ranging from A to B.  You must supply the values A and B.
    The end points A and B will be included in the set of
    possible random integers.  The resulting random integers
    may be used to select a simple random sample from a
    population, or may be studied (if saved) using another
    program such as ANALYZ or BOXPLOT or GOODFT.

    What is the lower limit A? 1
    What is the upper limit B? 10

    Note: For a Uniform Model:

    Theoretical skewness = 0.000 (Pearson coefficient)
    Theoretical kurtosis = 1.800 (Pearson coefficient)


                50 RANDOM INTEGERS FROM 1 TO 10

            4   10    3    9    5    8    1    1    7    3
            4    8    1    4    9    6    9    3    5   10
            6    2   10   10    4    9    4    6    4    9
            2    1    5    1    8    2   10    7    5    5
            5    4    5    6    2    6    3    6    4    4
```

```
OPTION TO SAVE THE DATA

Do you want to save these random numbers (y/n)? y
Which drive for saving (A, B, C, or D)? b
File name for saving ? unif33
File B:UNIF33 successfully saved.

Where now:   E=EXPLORE menu    R=Run RAND again    X=exit ? e
```

■ Case Study 19.3: Normal Random Numbers

As an experiment, we will generate fifty random normal numbers with a mean of 0 and a standard deviation of 1. Such a sample should be like the *standard normal model* ($\mu = 0$ and $\sigma = 1$). The standard normal model is the one described in Appendixes A and B. The theoretical probabilities of the outcomes are shown in Table 19.1, along with the expected frequencies. The actual frequencies differ somewhat from theory.

In actuality, the sample data look like Figure 19.3. The sample histogram differs noticeably from the ideal bell-shaped curve but is the difference significant? After all, we expect samples to vary somewhat. We would have to run the program GOODFT (not shown here) to obtain a precise answer, using the RAND-generated data file NORM33. It happens in this case that the sample is within the range that might be expected by chance, despite the somewhat ragged appearance of the sample histogram.

For the model, we specified the intended population parameters as

Population mean $= 0$

Population standard deviation $= 1$.

Table 19.1   Expected versus Actual Sample Frequencies

Range of X	Normal Probability	Expected Frequency	Actual Frequency
Under −3	.00135	.0675	0
−3 to −2	.02140	1.0700	0
−2 to −1	.13591	6.7955	10
−1 to 0	.34134	17.0670	13
0 to 1	.34134	17.0670	22
1 to 2	.13591	6.7955	4
2 to 3	.02140	1.0700	1
3 and over	.00135	.0675	0
Total:	1.00000	50.0000	50

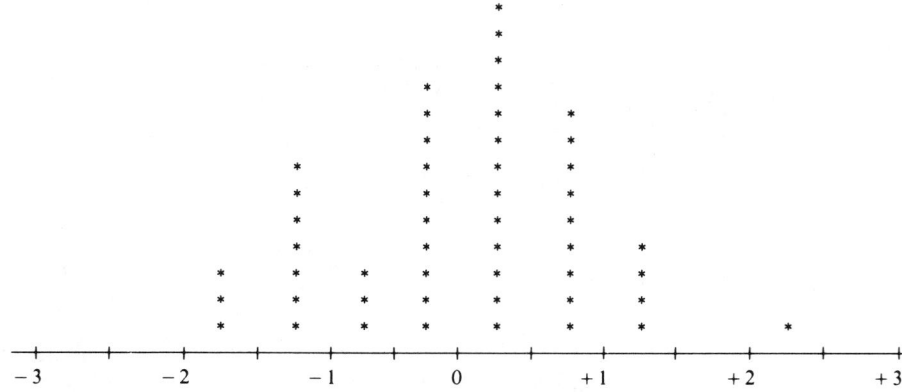

Figure 19.3   Actual Sample of Size n = 50

By running ANALYZ or BOXPLOT (printouts not shown) or by hand calculation we find that the sample mean is very close to the intended value, though the sample standard deviation differs somewhat from what we wanted:

$$Sample \ mean = -.063$$

$$Sample \ standard \ deviation = .865.$$

What about the quartiles? Using Appendix A or B, we find that for a standard normal distribution the quartile points should be

$$First \ quartile = -.674$$

$$Second \ quartile = .000$$

$$Third \ quartile = +.674.$$

Using Appendix B, you can verify that approximately 25 percent of the normal area lies to the right of $z = .674$ (and, of course, an equal area to the left of $z = -.674$). The second quartile is the median, which should be the same as the mean (zero in our example). In other words, half the data points should be above the mean and half below. The sample statistics are

$$First \ quartile = -.692$$

$$Second \ quartile = -.009$$

$$Third \ quartile = +.523.$$

While not exactly what we expected, these sample statistics seem fairly close to the intended values.

Finally, any normal distribution should be symmetric and mesokurtic in shape. That is, we expect to observe something close to

$$Skewness = 0$$

$$Kurtosis = 3.$$

For our sample data (using ANALYZ on file NORM33) we find that

Skewness = .064

Kurtosis = 3.107.

The sample skewness statistic suggests almost perfect symmetry (skewness close to zero), while the kurtosis statistic suggests a close approximation to a mesokurtic distribution (see Appendixes F and G for details of these comparisons).

```
                            RAND

        This program will generate up to 1000 random numbers
and save them in a file, if you wish.  The options are:

        1. Normal
        2. Uniform
        3. Poisson

Which option do you want (1, 2, or 3)? 1
How many random numbers do you want? 50
For random start, enter any 2-digit number: 33

PROCEDURE TO MAKE  50 NORMAL RANDOM NUMBERS

This program will make 50 normally-distributed random
decimal numbers with a chosen mean and standard deviation.
The resulting values, if saved, may be studied using
another program such as ANALYZ or BOXPLOT or GOODFT.

What is the desired mean? 0
What is the desired standard deviation? 1

Note: For a Normal Model:

Theoretical skewness = 0.000 (Pearson coefficient)
Theoretical kurtosis = 3.000 (Pearson coefficient)


        50 RANDOM NORMAL NUMBERS WITH MEAN = 0 AND S.D. = 1

   -0.298     0.839    -1.292    -0.229    -1.660     0.008     0.095
    0.078     1.035     0.074    -1.219     0.649     0.521     1.021
    0.001     0.969     0.532    -0.166    -1.012    -1.479     0.804
   -1.504    -0.022    -0.210     0.070    -1.007    -0.796    -1.740
   -0.935    -0.692    -1.099     0.416     0.523     0.426     0.304
   -0.257     1.179     0.332    -1.354    -0.277    -0.234     0.009
    0.162     0.843     1.020     2.443     0.087     0.586    -0.326
   -0.384


OPTION TO SAVE THE DATA

Do you want to save these random numbers (y/n)? y
Which drive for saving (A, B, C, or D)? b
File name for saving ? norm33
File B:NORM33 successfully saved.

Where now:  E=EXPLORE menu   R=Run RAND again   X=exit ? e
```

Case Study 19.4: Poisson Random Numbers

To generate a sample of Poisson random numbers, we run RAND using Option 3. We generate a sample of size n = 100 using a mean of μ = 2.5. According to theory, we should get the distribution shown in Table 19.2. The theoretical probabilities in Table 19.2 were obtained by running POIS with μ = 2.5.

As Table 19.3 shows, the actual sample contains too few 4s and perhaps too many 5s but is otherwise in fairly good agreement with the theoretical model shown in Table 19.2.

According to statistical theory, the sample mean and standard deviation of this Poisson model should be

$$\text{Mean} = \mu = 2.5$$
$$\text{S.D.} = \sqrt{\mu} = \sqrt{2.5} = 1.581.$$

The actual mean and standard deviation of the sample (calculated by hand or by using ANALYZ, BOXPLOT, or GOODFT with the saved data file POIS33) are

$$\text{Mean} = 2.590$$
$$\text{S.D.} = 1.609.$$

The sample mean and standard deviation are very close to the desired values. Moreover, the sample is positively skewed and leptokurtic, as any Poisson should be. This tends to confirm that RAND is working properly.

Similarly, with μ = 2.5 the theoretical shape parameters are

$$\text{Skewness} = 1/\sqrt{\mu} = 1/\sqrt{2.5} = .632$$
$$\text{Kurtosis} = 3 + 1/\mu = 3 + 1/2.5 = 3.400.$$

Table 19.2   Ideal Poisson Sample of Size n = 100

X	P(X)	Expected Frequency
0	.0821	8.21
1	.2052	20.52
2	.2565	25.65
3	.2138	21.38
4	.1336	13.36
5	.0668	6.68
6	.0278	2.78
7 or more	.0142	1.42
Total:	1.0000	100.00

Table 19.3   Actual Poisson Sample of Size n = 100

X	Frequency
0	5
1	22
2	28
3	22
4	8
5	9
6	4
7	2
Total:	100

The actual sample skewness and kurtosis statistics (from ANALYZ, BOX-PLOT, or GOODFT) are

$$Skewness = .729$$

$$Kurtosis = 3.051.$$

Thus we may say that our sample is slightly flatter than expected, but its skewness is very close to the theoretical value.

Finally, GOODFT demonstrates that file POIS33's fit to a Poisson model is acceptable. For details of this test, see GOODFT's Case Study 11.5, which is based on the file POIS33.

```
                    RAND

        This program will generate up to 1000 random numbers
and save them in a file, if you wish.  The options are:

    1. Normal
    2. Uniform
    3. Poisson

Which option do you want (1, 2, or 3)? 3
How many random numbers do you want? 100
For random start, enter any 2-digit number: 33


PROCEDURE TO MAKE  100 POISSON RANDOM NUMBERS

This program will make 100 Poisson-distributed random
integers with a chosen mean.  Mean cannot exceed 75.
The resulting values, if saved, may be studied using
another program such as ANALYZ or BOXPLOT or GOODFT.

What is the desired mean? 2.5
```

```
Note: For your Poisson Model:

Theoretical skewness = 0.632 (Pearson coefficient)
Theoretical kurtosis = 3.400 (Pearson coefficient)

          100 RANDOM POISSON INTEGERS WITH MEAN = 2.5
```

1	1	1	1	3	2	3	5	4	3
2	1	2	1	2	2	5	2	1	0
2	2	6	1	4	1	2	3	1	1
1	1	3	1	2	6	2	3	5	6
1	4	2	3	3	2	3	5	4	3
4	1	4	2	6	1	5	5	3	3
7	3	3	0	1	5	2	2	0	2
4	2	3	2	1	0	2	1	5	7
2	3	3	1	2	4	2	5	3	3
3	1	3	2	2	2	2	2	0	3

```
OPTION TO SAVE THE DATA

Do you want to save these random numbers (y/n)? y
Which drive for saving (A, B, C, or D)? b
File name for saving ? pois33
File B:POIS33 successfully saved.

Where now:  E=EXPLORE menu   R=Run RAND again   X=exit ? e
```

■ Technical Notes for Advanced Learners

Determined nonprogrammers may skip this section, which briefly describes the inner workings of program RAND. To make a uniformly distributed random integer, RAND uses the built-in BASIC random number generator RND as follows:

$$X = INT(A + RND*(B-A+1))$$

where B and A are the desired upper and lower limits. To see why this formula works, try making a sketch of how it maps RND into X (recall that RND is a uniformly distributed continuous variable between 0 and 1). The BASIC random number generator RND has been shown to have some flaws, but it suffices to illustrate the technique.

To make normally distributed random variate, RAND uses this algorithm:

$$K = 24$$
$$S = 0$$
$$FOR\ J = 1\ to\ K$$
$$S = S + RND$$
$$NEXT\ J$$
$$S = S - (K/2) / SQR(K/12)$$
$$X = MU + S*SIGMA$$
$$PRINT\ X$$

where MU is the desired mean and SIGMA is the desired standard deviation. The accuracy increases with K. For some purposes, K = 12 will suffice, but RAND uses K = 24. Statisticians generally use more elaborate methods for theoretical applications. However, this one gives acceptable results for learning about random numbers.

To generate Poisson random numbers, RAND finds the cumulative Poisson distribution F(X) for the desired mean and then uses this algorithm:

$$X = 0$$
$$1 \quad Y = RND$$
$$IF \ Y < F(X) \ THEN \ 2$$
$$X = X + 1$$
$$GO \ TO \ 1$$
$$2 \quad PRINT \ X$$

■ Exercises

19.1   Choose any limits A and B. Use RAND to construct a sample of twenty-five uniformly distributed random integers, and save them in a file called UNIF25. Using the same A and B, repeat this experiment using a sample size of one hundred, and save the results in a file named UNIF100.

19.2   Use ANALYZ to examine the data files UNIF25 and UNIF100. Make a printed copy of the results.

19.3   Use BOXPLOT to examine the data files UNIF25 and UNIF100. Make a printed copy of the results.

19.4   Based on your ANALYZ and/or BOXPLOT runs, do the two files appear close enough to a uniform distribution that you are comfortable concluding that the random number generator is working correctly? What is the effect of having a larger sample size? Discuss fully.

19.5   Choose any mean and standard deviation (μ and σ). Use RAND to construct a sample of twenty-five normally distributed random numbers, and save them in a file called NORM25. Using the same mean and standard deviation, repeat this experiment using a sample size of one hundred, and save the results in a file named NORM100.

19.6   Use ANALYZ to examine the data files NORM25 and NORM100. Make a printed copy of the results.

19.7   Use BOXPLOT to examine the data files NORM25 and NORM100. Make a printed copy of the results.

19.8   On the basis of your ANALYZ and/or BOXPLOT runs, do the two files appear close enough to a normal distribution that you are comfortable concluding that the random number generator is working correctly? What is the effect of having a larger sample size? Discuss fully.

19.9   Choose any mean (μ). Use RAND to construct a sample of twenty-five Poisson random numbers, and save them in a file called POIS25. Using the same mean, repeat this experiment using a sample size of one hundred, and save the results in a file named POIS100.

19.10 Use ANALYZ to examine the data files POIS25 and POIS100. Make a printed copy of the results.

19.11 Use BOXPLOT to examine the data files POIS25 and POIS100. Make a printed copy of the results.

19.12 On the basis of your ANALYZ and BOXPLOT runs, do the two files appear close enough to a Poisson distribution that you are comfortable concluding that the random number generator is working correctly? What is the effect of having a larger sample size? Discuss fully.

Optional Exercises for Ambitious Learners

19.13 Use GOODFT to test your files UNIF25 and UNIF100 for goodness of fit to a uniform distribution. Discuss the results fully. Refer to Appendixes F and G to help you discuss the shape of the samples.

19.14 Use GOODFT to test your files NORM25 and NORM100 for goodness of fit to a normal distribution. Discuss the results fully. Refer to Appendixes F and G to help you discuss the shape of the samples.

19.15 Use GOODFT to test your files POIS25 and POIS100 for goodness of fit to a Poisson distribution. Discuss the results fully. Refer to Appendixes F and G to help you discuss the shape of the samples.

■ References

1. Chambers, John M., *Computational Methods for Data Analysis* (New York: John Wiley and Sons, Inc., 1977).

2. Dewdney, A. K., "Computer Recreations," *Scientific American,* Vol. 253, 1985, pp. 17–20.

3. Hastings, N. A. J. and Peacock, J. B., *Statistical Distributions* (New York: John Wiley and Sons, 1975).

4. Knuth, Donald E., *The Art of Computer Programming* (Reading, Mass.: Addison-Wesley Publishing Company, 1969), pp. 102–113.

20

SAMPLER/Sampling and Confidence Intervals

Introduction

The Central Limit Theorem says that if random samples of size n are drawn from a population with mean μ and standard deviation σ, the sample means will approach a normal distribution with mean μ and standard deviation $\sigma/\sqrt{n}$ as n increases. Thus, we should be able to use the areas of the normal probability density function to construct confidence intervals for the population mean, which will enclose the true mean a given percentage of the time.

However, the theory outlined in the preceding paragraph may seem a bit abstract. Using SAMPLER, you can actually draw samples of a desired size from a population of your choice, either with or without replacement. For each sample, SAMPLER will calculate the mean, standard deviation, skewness, kurtosis, and 95 percent confidence limits for the true population mean μ. Then it will compute the mean and standard deviation of all the means so they can be compared with the theoretical values derived from the population.

Using SAMPLER, you can empirically explore the laws of statistics to see whether 5% of the confidence intervals really fail to enclose the population mean μ. You can try various kinds of populations, including some that are symmetric and well-behaved or others that are skewed and bizarre. If the population is well-behaved, the laws of statistics are easily illustrated. But be forewarned that when you are dealing with bizarre populations with outliers, sample variation can be extreme.

In sampling without replacement, strange things begin to happen if you try to sample a very large fraction of the population. A moment's reflection will reveal why the program may slow down, especially if the population is large. But SAMPLER will warn you if it is getting into such a situation, so no special caution is needed.

◼ Case Study 20.1: Sampling NYSE Bond Prices

Table 20.1 shows the prices of eighty bonds listed on the New York Stock Exchange (NYSE), representing every twentieth bond listed on the stock exchange (excluding convertible bonds or companies in receivership). For each

Table 20.1   Yields and Prices of Randomly Selected NYSE Bonds

Firm	Coupon	Mature	Price	Firm	Coupon	Mature	Price
AlaP	8.75	2007	94.00	KaufB	12.25	1999	103.00
AlldC	5.20	1991	86.75	LearS	10.00	2004	101.13
Alcoa	9.00	1995	100.13	LouGs	9.25	2000	101.00
ACyan	7.38	2001	84.00	MarCor	6.50	1988	88.50
Ames	10.00	1995	100.13	MerL	11.68	1987	104.25
AshO	6.15	1992	88.88	Mobil	8.50	2001	94.13
AvcoC	7.50	1993	92.50	MtSTl	9.68	2015	101.75
Banka	7.88	2003	83.25	NtMed	12.75	2000	106.88
BellCn	9.00	2008	98.00	Navstr	14.50	2001	114.75
Blair	13.68	1998	101.88	NiMp	8.35	2007	86.68
BurNo	8.60	1999	99.88	NwnBl	7.50	2005	88.13
CaroT	7.75	2001	89.50	OhBIT	9.00	2018	97.88
ChNY	8.40	1999	97.00	Ownlll	7.68	2001	87.00
ChvrnC	11.00	1990	107.25	PGE	8.00	2003	90.00
ChryF	9.80	1988	103.50	PSwAir	6.00	1987	95.00
CitSv	6.13	1997	72.00	PAA	11.25	1986	100.00
ColuG	7.50	1997	88.25	Petrie	8.00	2010	133.00
CmwE	8.00	2003	88.50	PhilEl	11.00	2000	101.00
ConEd	4.75	1990	90.13	PhillP	12.88	1992	103.00
CnPw	5.88	1996	73.25	PotEl	11.88	1989	106.50
CrnPd	5.75	1992	88.25	PSNH	15.75	1988	106.68
CritAc	11.25	2015	101.25	RJR	7.38	2001	88.00
DetEd	6.00	1996	81.50	Revlin	11.75	1995	100.25
Dow	8.68	2008	93.00	SanD	10.00	2006	102.25
DugL	5.00	2010	59.75	SearA	8.38	1986	100.00
Ens	9.75	1995	99.13	Socny	4.25	1993	80.00
FinCpA	6.00	1988	90.00	Spiegl	5.00	1987	97.50
FroC	8.50	1991	101.50	Subne	5.50	1992	81.00
Fuqua	7.00	1988	97.00	Tenco	14.50	2006	111.50
GEICr	11.75	2005	108.13	TexC	13.68	1994	109.00
GMA	8.75	2000	99.13	TxOG	16.68	1991	110.75
GMA	11.75	1989	105.00	TWA	4.00	1992	59.50
Gene	15.25	1994	99.00	TCFox	13.25	2000	104.68
GaPw	16.13	2011	112.50	UnEl	10.50	2005	104.00
GtNor	2.68	2010	40.00	USBO	7.75	2002	85.00
GlfRes	12.50	2004	99.75	ValerNG	16.75	2002	112.00
Honey	9.38	2009	100.00	WellF	8.60	2002	97.00
ITTF	8.10	1992	97.00	WUTI	9.25	1997	75.00
IngR	8.05	2004	88.00	Woolw	7.38	1996	95.00
IntTT	8.90	1995	100.68	Zapt	10.25	1997	42.00

Source: *New York Times,* July 20, 1986, pp. 12F–24F. Caution: Data are for classroom use only and are not for investment decisions.

bond we record the coupon rate (or original interest rate), year of maturity, and closing price on Friday, July 18, 1986. The price is expressed as a percent of face value (par), so a bond listed at 100 is selling at face value (those below 100 are said to be selling at a discount, while those above 100 are selling at a premium).

Using the EXPLORE editor, we enter the column of bond prices into a file named PRICES, which becomes the population (N = 80) that we will sample. When sampling with replacement, each 95 percent confidence interval will have the following form:

$$\overline{X} \pm 1.96\sigma/\sqrt{n}.$$

In this formula, $\overline{X}$ is the sample mean, $Z = 1.96$ is the normal variate value to include 95 percent of the area, σ is the population standard deviation of the file we are sampling, and n is the sample size.

If we are sampling without replacement, SAMPLER will narrow the interval by multiplying by the finite population correction factor (FPCF):

$$\text{FPCF} = \sqrt{\frac{N - n}{N - 1}}$$

where N is the population size. Therefore, the form of a finite-population confidence interval is

$$\overline{X} \pm Z\sigma/\sqrt{n} \sqrt{\frac{N - n}{N - 1}}.$$

If n = N (the largest possible sample), the FPCF will be zero, yielding a zero-width interval. At the other extreme, if n = 1 (the smallest possible sample), the FPCF will be unity. Remember that the FPCF is used by SAMPLER only if you are sampling without replacement.

To illustrate, let us take ten samples of size n = 16 with replacement. The SAMPLER printout is shown at the end of the chapter. In our experiment, all ten sample confidence bands enclose the true population mean (μ = 94.55049). Ten out of ten is about as close to 95 percent as we can get. The mean in sample 8 is too high, while the means in samples 7 and 9 are too low. But overall the sample means are fairly close to the population mean, and the mean of the ten sample means is close to μ:

True population mean: μ = 94.55

Mean of 10 sample means: $\overline{\overline{x}}$ = 94.19

Since we are sampling with replacement, the theoretical standard deviation of the ten sample means is $\sigma/\sqrt{n}$ = 14.34347/$\sqrt{16}$ = 3.59, which is close to the standard deviation of the ten sample means:

Theoretical S.D.:   3.59

S.D. of 10 sample means: 3.26.

However, we may have been lucky. The printout shows a rather erratic standard deviation. (Samples 1, 7, and 9 have larger standard deviations than the population, while the standard deviation is much too small in samples 5 and 10.) This suggests that we may be picking up outliers in some samples. From ANALYZ (printout not shown) we discover that the eighty observations in the file PRICES are skewed by two outliers in the left tail:

$$\text{GtNor at } 40.00 \; (Z = -3.779),$$
$$\text{Zapt at } 42.00 \; (Z = -3.641).$$

We also have two near-outliers in the left tail:

$$\text{TWA at } 59.50 \; (Z = -2.428),$$
$$\text{DugL at } 59.75 \; (Z = -2.411).$$

These bonds are far below the mean price of 94.55 (perhaps a little library research and Case Study 17.1 can tell you why). At the other end of the scale, in the right tail, only one observation lies at least two standard deviations above the mean:

$$\text{Petrie at } 133.00 \; (Z = 2.664).$$

The rest of the population values are not far from the mean. If any of these five extreme values happens to be chosen in a given sample, it can have a major impact, since the sample size is only n = 16. This helps to explain why our standard deviations are erratic. If two or more extreme values offset one another, the sample mean can still be close to the population mean (as in sample 2), even though the standard deviation is too large.

What about shape? In nine of our ten samples a leptokurtic population is indicated, as would be expected since the population is leptokurtic (kurtosis = 6.849). This indicates a relatively sharp "peak" in the histogram. In nine of our ten samples the skewness coefficient has a negative sign, indicating left-skewness, as we would expect from the left-skewed population (skewness = -1.387). Even in sample 5 the skewness coefficient does not differ greatly from zero. Thus from most of the samples we would correctly infer that the histogram of the population looks rather like Figure 20.1.

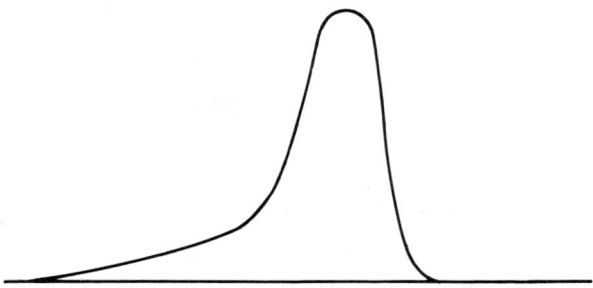

Figure 20.1 Population Shape

You should refer to Appendixes F and G for the limits of the sample skewness and kurtosis coefficients. However, our sample of n = 16 is a bit too small to use them (we kept it small so that it would be easier to explain). One final caution: When you are sampling without replacement, Appendixes F and G will yield wider intervals than are appropriate.

```
                          SAMPLER

        This program will choose repeated random samples of
n items from a saved data file of N items, and will compute
summary statistics for each sample.  The sample statistics
may then be compared with the population, to see whether
the laws of statistics are borne out.  Sampling may be with
or without replacement.  The maximum size of the file to be
sampled is 1000 items.  You may take up to 100 samples.

Name of file to be sampled? prices


        SETTING UP SAMPLE CHARACTERISTICS

File to be sampled    = PRICES
Population file size = 80    items
Size of each sample   = 16
Number of samples     = 10
Sample with replacement (y/n)? y
Input a 1-digit random integer to start? 3
```

```
        SAMPLES OF 16 ITEMS FROM POPULATION OF 80 ITEMS (File = PRICES)
        ================================================================
                                                      95% CI for Mean
Sample      Mean      St. Dev.   Skewness   Kurtosis   Lower      Upper
        ----------------------------------------------------------------
    1     90.06313    18.60110    -1.518     4.476    83.03483    97.09142
    2     95.22250    16.10094    -2.679     9.942    88.19420   102.25079
    3     95.38750    16.21143    -2.323     8.597    88.35920   102.41579
    4     94.45750    11.08632    -1.996     7.252    87.42921   101.48580
    5     97.50875     7.16013     0.385     3.455    90.48045   104.53705
    6     94.27500    15.45261    -1.317     3.869    87.24670   101.30330
    7     89.87563    18.21345    -1.425     4.875    82.84733    96.90392
    8     98.72750    11.31023    -0.635     2.756    91.69920   105.75580
    9     89.73563    19.44391    -1.311     3.604    82.70733    96.76392
   10     96.63625     9.60466    -0.530     3.693    89.60796   103.66455
        ----------------------------------------------------------------
True Pop 94.55049    14.34347    -1.387     6.849

Mean of X̄ = 94.18893

S.D. of X̄ = 3.260068

Where now:  E=EXPLORE menu    R=Run SAMPLER again    X=exit ? e
```

■ Exercises

20.1 Choose one of the numerical columns in Table 20.1 as your population, using the EXPLORE editor to enter the data into a file (suggested file names are COUPON, MATURITY, and PRICES. Run ANALYZ on your file, and print the screens showing the standardized values and histogram. Note any unusually high or low data values which might adversely affect confidence intervals for μ. Write a short paragraph describing your population's central tendency, dispersion, and shape.

20.2 Run SAMPLER three times, using sample sizes of 4, 16, and 64. In each case, take ten samples. Print each screen for analysis.

20.3 Find the percentage of the confidence intervals that enclose the true population mean from Exercise 20.1. If it is not exactly 95 percent, can you suggest why not? Plot each confidence interval on a graph, showing the true mean on a scale like this:

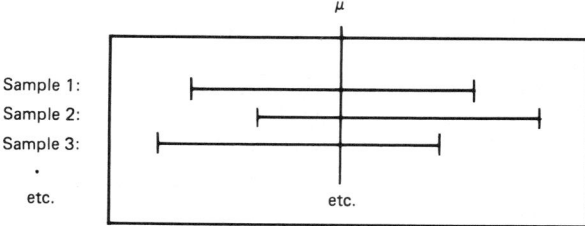

Discuss what your graph shows. Can you see at a glance which confidence intervals almost failed to include μ? Are outliers a possible cause?

20.4 How close were your samples to the theoretical mean and standard deviation of the means? Is the difference within chance?

20.5 Examine the skewness and kurtosis in each of your three SAMPLER runs. Do the samples conform to the population's shape? (Refer to Appendixes F and G). Discuss anything unusual in your results.

20.6 Repeat Exercise 20.2, but sampling without replacement. How much difference does the finite correction make in the confidence interval width?

20.7 Take 100 samples of size 4. Do not save the screens, but just look at each screen and count the number of times your confidence intervals fail to include the true population mean. Discuss your results. Increase the sample size to 25 and describe the results of that experiment.

20.8 Repeat Exercises 20.1–20.8 using any of the STATE database files. The label file for ANALYZ, if desired, will be the state names (file STATE).

20.9 Using a population file of your choice, take ten samples of size n = 1 without replacement. What is the width of the confidence intervals? Now take ten more samples without replacement, this time letting n = .9N. What happens to the width of the confidence intervals? Why? Does the program slow down? If so, why?

SAMSIZE/Sample Size for Interval Estimates

Introduction

"How large a sample should I take?" is a very common statistical question asked by business decision makers. The answer, as you might expect, depends on the situation. To determine the required sample size, a consulting statistician would need to ask several specific questions

What are you trying to estimate?

Do you have any prior information about it?

What precision (interval width) is required for your estimate?

What confidence level is required?

What is the population size?

There are many correct answers to the sample size question, depending upon the tradeoffs you are willing to make. When you have a fixed budget for sampling, you can obtain greater precision (narrower interval) if you are willing to accept a lower confidence level, or vice versa. Sometimes, the only way to stay within budget is to reduce both precision and confidence.

Sample size calculations can clarify the alternatives that are available to the decision maker. The formulas are time-consuming enough that a decision maker might be discouraged from exploring all the alternatives unless a computer program like SAMSIZE is available. Program SAMSIZE makes it easy to try various options and to prepare summary tables describing the problem, without getting bogged down in tedious algebra.

If we want to construct interval estimates whose half-width is e (the error we will accept), the intervals will have the form

$$\overline{X} \pm e \quad \text{(for a mean)}$$

or

$$\hat{p} \pm e \quad \text{(for a proportion)}.$$

The formulas used by SAMSIZE to obtain the required sample size (n_0) are as follows:

$$\text{For a mean:} \quad n_0 = [Z\sigma/e]^2;$$
$$\text{For a proportion:} \quad n_0 = [Z/e]^2 p(1 - p).$$

where

Z = normal distribution value for the desired confidence level,

σ = the population standard deviation (usually estimated by s from a preliminary sample),

e = the error (precision) that we will accept in our interval,

p = the assumed population proportion (use .5 if you have no prior information).

The above formulas give a final result if we are sampling an infinite population or if sampling is with replacement. If the population is finite, we obtain the *adjusted sample size* (n) using the formula

$$n = n_0 N/[n_0 + (N - 1)].$$

This formula includes the *finite population correction,* which requires knowing the population size (N). These formulas, along with the required z values (shown in Table 21.1) are built into program SAMSIZE. All you have to enter is the desired error (e), parameter estimate (σ or p), and population size (N).

Case Study 21.1: Sample Size for Mean Executive Compensation

Table 21.2 shows a sample of nineteen chief executive officers (CEOs), chosen at random from a list of the 793 largest companies in the United States. The sampling methodology was to choose the tenth company on the top line of each page in the nineteen-page list. We record the company name, and its CEO's name, age, total compensation (thousands of dollars, including bonuses and stock gains), years of experience in the company, and type of business background.

From this sample we could construct many kinds of estimates. For example, we could estimate the mean compensation (or age or experience) of all

Table 21.1   Z Values for Various Common Confidence Levels

Confidence Level	80%	90%	95%	98%	99%
Z Value	1.282	1.645	1.960	2.326	2.576

Table 21.2   Profile of Randomly Chosen CEOs in Large U.S. Firms

Firm	Chief Executive	Age	Compensation	Years at Firm	Background
Alexander & Baldwin	R. J. Pfeiffer	66	1327	26	Operations
AmeriTrust	J. V. Jarrett	54	771	12	Banking
Bell Atlantic	T. E. Bolger	58	909	43	Admin.
Centex	P. R. Seegers	56	466	24	Finance
Comdisco	K. N. Pontikes	46	1409	16	Marketing
Digital Equip.	K. H. Olsen	60	755	28	Founder
Fin. Corp. Santa Barb.	P. R. Brinkerhoff	43	808	1	Finance
Fourth Financial	J. L. Haines	59	297	24	Admin.
GTE	T. F. Brophy	63	1056	28	Legal
IBM	J. F. Akers	51	736	25	Marketing
Louisiana Banc-shares	C. W. McCoy	66	220	26	Banking
Mesa Petroleum	T. B. Pickens, Jr.	58	9877	29	Founder
N. Amer. Philips	C. Bruynes	53	864	11	Marketing
Pfizer	E. T. Pratt, Jr.	59	2260	21	Finance
Roadway Services	C. F. Zodrow	63	508	27	Finance
Squibb	R. M. Furlaud	63	2780	30	Legal
Thrifty Corp.	L. H. Straus	71	1118	40	Legal
USF & G	J. Moseley	55	420	32	Insurance
Yellow Freight Sys.	G. E. Powell, Jr.	60	662	34	Admin.

Compensation is in thousands of dollars.
Source: *Forbes,* Vol. 137, No. 12, June 2, 1986, pp. 152–188.

CEOs in the population. Let's illustrate how SAMSIZE works for a mean, using our preliminary sample of n = 19 (from a population of N = 793 firms). Our preliminary sample yields the sample means and standard deviations shown in Table 21.3.

Table 21.3   Sample Statistics for Nineteen CEOs

Statistic	Age	Compensation	Experience
Sample Mean	58.105	1433.84	25.105
Sample S.D.	6.887	2142.15	9.955

Suppose we would like to make confidence intervals for the true mean compensation of all CEOs, with these widths (chosen arbitrarily):

$$\overline{X} \pm 400 \quad \leftarrow \quad \text{large error (low precision)},$$
$$\overline{X} \pm 200 \quad \leftarrow \quad \text{medium error (medium precision)},$$
$$\overline{X} \pm 100 \quad \leftarrow \quad \text{small error (high precision)}.$$

To obtain the required sample sizes, we first run SAMSIZE using $\sigma = 2142$, $N = 793$, and $e = 400$ (see the sample printout). Then we run SAMSIZE again using $e = 200$, and again using $e = 100$. The results are summarized in Table 21.4.

Since our population is finite, the "Finite" rows of the table are the relevant ones, but it is instructive to see how much larger the required samples would be without the finite population correction, particularly at high confidence levels and low desired error (e). As would be expected, the required sample size increases as we reduce the desired error. Also, higher confidence levels require a larger sample size. When the desired error is small (say, $e = 100$) and the desired confidence level is high (say 99 percent), we might just as well study the entire population, since the sample sizes are not far from the population size ($N = 793$).

Now, let's turn the problem around. Suppose you want to take a sample of about 150 items (because of a time or budget constraint). You could use 99 percent confidence and an error of 400. Or you could use 80 percent confidence with an error of only 200. There are other alternatives in between. But you could not achieve an error of 100 without taking a sample larger than 150 (at least using the confidence levels shown).

Only the decision maker can make these tradeoffs, which are not based on statistical grounds. But it is a proper responsibility of the statistician to explain to a client that money could be saved by sacrificing some accuracy or confidence. Business people often start off with a wish for extreme accuracy

Table 21.4   Required Sample Sizes for Estimating μ if $\sigma = 2142$

Run	Population Treated As	e	σ	Confidence Level 80%	90%	95%	98%	99%
1	Finite	400	2142	45	71	97	130	154
	Infinite	400	2142	48	78	111	156	191
2	Finite	200	2142	153	224	284	349	389
	Infinite	200	2142	189	311	441	621	762
3	Finite	100	2142	387	485	548	602	630
	Infinite	100	2142	755	1242	1763	2483	3045

```
                          SAMSIZE

This program calculates the required sample size for common
levels of confidence and allowable error (e), either from
a finite or infinite population, for two types of estimates:

     1. Confidence interval for a mean:      X̄ +/- e
     2. Confidence interval for a proportion:  p +/- e

Which do you want (1 or 2)? 1
Allowable error e (use same units as X)? 400
Standard deviation (or prior estimate)? 2142
Sampling without replacement is assumed ...
Finite population (y/n)? y
Population size (integer)? 793


      SAMPLE SIZE REQUIRED FOR CONFIDENCE INTERVALS FOR A MEAN
      =========================================================
                            Confidence Level
                    80%      90%      95%      98%      99%
      -----------------------------------------------------
Sample size assuming
finite population:   45       71       97      130      154

Sample size assuming
infinite population: 48       78      111      156      191

Z-Value:           1.282    1.645    1.960    2.326    2.576
      -----------------------------------------------------

Allowable error is plus or minus 400
Assumed population standard deviation is 2142
Population size is 793 items.
Sample size is rounded to next higher integer.

Where now:   E=EXPLORE menu    R=Run SAMSIZE again    X=exit ? e
```

and 99 percent confidence but change their minds when they find out how much it will cost.

To use the formula for a mean, we need a preliminary sample to estimate σ. It may seem circular that we must take a sample to estimate the size of the sample that we plan to take. But a preliminary sample is usually a good idea, to give us a feeling for the data and to give us some idea what to expect.

In assessing our SAMSIZE calculations a statistician might worry about plugging in $\sigma = 2142$, which may be inaccurate. Not only is our preliminary sample rather small, but also we have one possible outlier (T. B. Pickens, Jr., of Mesa Petroleum) whose compensation is about four times larger than that of the next chief executive. The mean and standard deviation are probably inflated as a result. A larger preliminary sample might yield a better estimate of σ, but ever-larger preliminary samples tend to defeat the purpose of SAMSIZE. When we plug in a sample result, we take a chance.

You could choose any desired error (not just 400, 200, and 100). The appropriate error varies with the problem. Here an error of 200 is about ±14 percent of the preliminary sample mean (1434). However, an error of 50 would lead to extremely large sample sizes (essentially equal to the entire population) because it would be about ±3 percent of the sample mean, (1434). Only trial-and-error can suggest an appropriate error level for a given data set.

In some applications you might need other confidence levels, but SAM-SIZE should cover most situations. A program to do these calculations is not difficult to write if you would like to experiment with different assumptions.

■ Case Study 21.2: Sample Size for a Proportion

Now let us use SAMSIZE to estimate the proportion of CEOs who have a background in finance. We might pause to consider that all chief executives acquire a strong exposure to finance in their climb to the top; the *Forbes* listing presumably reflects their primary corporate experience before becoming a CEO. It would be unreasonable to assume that the label reflects college training, although finance majors should have as good a chance as anyone at becoming CEOs. In this sample, four out of nineteen CEOs have a finance background, so the sample proportion is $\hat{p} = 4/19 = .21$. Now suppose we would like to construct confidence intervals with these widths (chosen arbitrarily):

$$\hat{p} \pm .10 \quad \leftarrow \quad \text{large error.}$$
$$\hat{p} \pm .05 \quad \leftarrow \quad \text{medium error.}$$
$$\hat{p} \pm .02 \quad \leftarrow \quad \text{small error.}$$

Since we do not know the true population proportion, we plug the sample estimate $\hat{p} = .21$ into program SAMSIZE, which we run three times to obtain the results shown in Table 21.5.

The required sample size increases as we reduce the allowable error (e) and as we increase the level of confidence. The infinite population sample sizes are irrelevant but do show the striking effect of the finite population correc-

Table 21.5   Required Sample Sizes for Estimating p Using $\hat{p}$ = .21

Run	Population Treated As	e	$\hat{p}$	Confidence Level				
				80%	90%	95%	98%	99%
1	Finite	.10	.21	27	43	60	81	97
	Infinite	.10	.21	28	45	64	90	111
2	Finite	.05	.21	96	147	194	248	284
	Infinite	.05	.21	110	180	255	360	441
3	Finite	.02	.21	367	465	530	587	616
	Infinite	.02	.21	682	1123	1594	2244	2753

tion formula. From this table we infer that to achieve a small error (say, e = .01), we would need such large sample sizes that we might just as well take a census of the entire population. Even e = .02 requires a sample usually exceeding half the population.

By using our sample estimate $\hat{p}$ = .21, could we have introduced an error into the calculation? Definitely so. After all, our sample size (n = 19) was rather tiny, particularly for estimating a sample proportion. A more conservative procedure would be to assume that p = .50. This amounts to assuming the worst case and will yield a sample size that is sure to be large enough (unless the normality condition itself is severely violated). The results of re-running SAMSIZE are shown in Table 21.6. Required sample sizes are invariably greater than if we assumed p = .21.

Should we always assume that p = .50? Usually, statisticians do assume that p = .50 because they want to be conservative. Also, there may be no preliminary sample. But the cost of being conservative is that the survey will take more time and may cost more than is necessary.

Also, we are not bound by the sample size that we select at this time. If the required sample size is 1000 (assuming p = .5), we can take the first 500 and see whether we are close enough, or we can revise our required sample size downward on the basis of the early data. We are not stuck with the assumption that p = .5, although it is a good *conservative* assumption.

How hard would it be to take a preliminary sample of one hundred CEOs and see what proportion have a finance background? Such a survey is easy in this case, since it merely involves looking at names on a page. Of the first one hundred names it is easily determined that thirteen have a finance background, so we could use $\hat{p}$ = .13 as a better estimate to refine the sample size calculations. As always, a preliminary survey is a good idea, just to get a feeling for the data.

Table 21.6 Required Sample Sizes for Estimating p Using p = .5

Run	Population Treated As	e	p	Confidence Level				
				80%	90%	95%	98%	99%
1	Finite	.10	.50	40	63	86	116	138
	Infinite	.10	.50	42	68	97	136	166
2	Finite	.05	.50	137	202	260	322	362
	Infinite	.05	.50	165	271	385	542	664
3	Finite	.02	.50	448	541	597	643	666
	Infinite	.02	.50	1028	1692	2402	3382	4148

```
                           SAMSIZE

       This program calculates the required sample size for common
       levels of confidence and allowable error (e), either from
       a finite or infinite population, for two types of estimates:

              1. Confidence interval for a mean:        X̄ +/- e
              2. Confidence interval for a proportion:  p +/- e

       Which do you want (1 or 2)? 2
       Allowable error  (0 < e < 1)? .10
       Estimated p  (0 < p < 1) or enter .5 if no idea? .21
       Samping without replacement is assumed ...
       Finite population (y/n)? y
       Population size (integer)? 793

       SAMPLE SIZE REQUIRED FOR CONFIDENCE INTERVALS FOR A PROPORTION
       ==============================================================
                                     Confidence Level
                              80%     90%     95%     98%     99%
       -------------------------------------------------------------
       Sample size assuming
       finite population:     27      43      60      81      97

       Sample size assuming
       infinite population:   28      45      64      90     111

       Z-Value:             1.282   1.645   1.960   2.326   2.576
       -------------------------------------------------------------

       Allowable error is plus or minus .1
       Assumed population proportion is .21
       Population size is 793 items.
       Sample size is rounded to next higher integer.

       Where now:  E=EXPLORE menu   R=Run SAMSIZE again   X=exit ? e
```

Exercises

21.1 Using the sample standard deviations shown in Table 21.3, what sample sizes would be required to estimate the true mean age of CEOs for the 793 largest companies in the United States with an error not exceeding five years? Two years? One year? Arrange your results in a table and discuss the effects of increasing confidence and decreasing error. Sketch a graph of sample size as a function of error. What is its shape? Do the same for sample size as a function of confidence.

21.2 How much difference would it make if you sampled with replacement or let the population be all U.S. companies? Would the same standard deviation still be appropriate? What alternatives could be used?

21.3 Repeat Exercises 21.1 and 21.2 using years of experience, and an error of six years, four years, and two years.

21.4 Using the preliminary data in Table 21.1 to make a preliminary estimate, what sample sizes would be required to estimate the true proportion of

CEOs for the 793 largest companies in the United States who have marketing background, with an error not exceeding .10? .05? .01? What error would you recommend, and why? Arrange your results in a table. Sketch a graph of sample size as a function of confidence and also as a function of error. Discuss the shape of these graphs. If you ignore the preliminary sample information and assume that p = .5, how much difference would it make in the sample size calculations?

21.5 A preliminary sample of 82 primitive art works found in the Americas indicates that 72 were crafted by right-handed artists. What sample sizes would be needed to estimate the true proportion of right-handed art works with an error not exceeding .10? .05? .01? Would these samples sizes actually be feasible? Why or why not? Is the finite population correction needed? What difference would it make if you assumed that p = .5 instead of using the available sample information? (Source: *Science,* Vol. 198, November 1977, pp. 631–632).

21.6 A poll of 1007 randomly chosen U.S. adults showed that 58 percent of the respondents chose the response "absolutely yes, without question" when asked the question "In your opinion, does a terminally ill person and/or one who is permanently bedridden and kept alive by medical machinery have the right to request to be allowed to die?" What sample size would be needed to ensure an error not exceeding .01? Might a different error be advisable? Explain your reply. (Source: *Detroit Free Press,* May 2, 1985, p. 19A).

21.7 A random sample was chosen by picking every fourth aircraft on a list of 33 aircraft. For each plane, the make and model, year of manufacture, and initial rate of climb (in feet per minute) were recorded. Rates of climb are measured at sea level, at standard temperature and air pressure. Use the sample data in Table 21.7 to find a preliminary estimate of the population standard deviation. Then use SAMSIZE to estimate the sample size required to yield an estimate of the mean initial rate of climb for all aircraft listed in the population (N = 33) with an error not exceeding 100 fpm. How large a sample would be required if the allowable error were decreased to 50 fpm? What sample size would you actually recommend?

21.8 A survey of 49 statistics students revealed that the mean number of children judged to be "ideal for a married couple" was 2.592 with a standard deviation of .911. How large a sample would be required to estimate the true mean with an error not exceeding .25 children? .15 children? Is a finite correction needed? With the existing preliminary sample, what approximate error and confidence could be achieved? (Source: confidential survey.)

21.9 A sample of six aerospace and defense contractors revealed a mean debt/equity ratio of 34.1 with a standard deviation of 31.3 (Lockheed 20.7, Teledyne, 92.4, Raytheon 4.3, Tracor 42.0, Sundstrand 30.1, Westinghouse 15.2). If there are 24 major aerospace and defense contractors on

Table 21.7 Initial Rate of Climb, Selected Aircraft

Aircraft	Year	Rate of Climb
Aeronca 7AC Champion	1946	370
Avions Mudry CAP 10B	1983	1000
Cessna 177A Cardinal	1969	760
Cessna TU 206G Turbo	1982	1010
Great Lakes 2T-1A-2	1982	1150
Luscombe Silvaire 8A	1946	800
Piper PA-22-150 Tri-Pacer	1956	750
Piper PA-28-181 Archer	1981	735
Taylorcraft F-21	1982	875

Source: *AOPA Pilot,* September 1984, p. 28.

the entire list, what sample size is needed to estimate the population mean with an error not exceeding 10? What advice would you have concerning the sample size that should be taken? (Source: *Forbes,* Vol. 137, No. 1, 1986, p. 58.)

Optional Exercises for Ambitious Students

21.10 Before passage of a ''bottle law'' a survey of road litter on 36 randomly chosen 1000-foot segments of highway reveals an average of 43.5 bottles or cans per segment. How large a sample would be required to ensure an error not exceeding 5 bottles/cans? 2 bottles/cans? (Hint: Since there is no standard deviation, assume that we have a Poisson event whose mean is 43.5, and use this mean to estimate the standard deviation, as explained in Chapter 17.) In a follow-up survey after a ''bottle law'' was passed, the mean had fallen to 5.51. Without doing any tests, do you think this difference is statistically significant? (Source: *Detroit Free Press,* July 20, 1981, p. 1A).

22 SORTER/Sorting
Several Variables

Introduction

Sorting a column of data will often tell the statistician all that is necessary. From a sorted column of numbers it is easy to construct histograms, find the median or quartiles, and assess the range of the data. It is also possible to spot outliers (unusually extreme data values).

SORTER will sort up to five data files in ascending order and will print the sorted columns side by side. Two options exist:

1. Sort each variable separately.
2. Sort all variables on one variable.

If you choose the second option, all the files are presumably related to a common unit of observation (for example, the fifty states in the United States). If you choose the first option, files of unequal length may be used. Either numeric or alphabetic data are acceptable under both options. The time required to sort data files will depend on the length of the files.

You may save the sorted data columns if you wish. The program SORTER will request new file names and will not alter the old files. If the new file names conflict with any existing files, a message will be printed so that you can avoid erasing old files. In SORTER, long character strings are truncated in the display, and numerical data are formatted as in LAYOUT.

Case Study 22.1: Sorting Nations by Population Density

Table 22.1 shows data for twelve nations of the world: annual birth rate per 1000 persons, gross national product per capita (converted to U.S. dollars at prevailing exchange rates), and population density (persons per square mile of land area).

Table 22.1   Statistical Profile of Twelve Nations

Country	Birth Rate	GNP per Person	People per Square Mile
Afghanistan	32	170	66
Bangladesh	47	110	1665
Bolivia	47	890	12
Chad	44	220	10
Egypt	38	540	110
Guatemala	43	1150	174
Indonesia	43	460	192
Norway	13	14930	33
Spain	16	5600	200
Tanzania	47	230	54
United States	15	11590	64
Vietnam	42	150	434

Source: *Reader's Digest 1982 Almanac* (Pleasantville, N.Y.: Reader's Digest Association, 1982), pp. 467–475.

Using the EXPLORE editor, we create four data files, each containing twelve observations:

COUNTRY = name of the nation,

B-RATE = annual birth rate per 1000 persons,

GNP-CAP = gross national product per person,

DENSITY = population density per square mile.

The following printout shows how all four variables can be sorted on DENSITY. By sorting the names of the nations too, it is easy to see which nations have highest population density. (Bangladesh and Vietnam) and which have the lowest (Chad and Bolivia).

The other option (sorting each variable separately) is not illustrated. The printout would look about the same, except that each variable would be sorted upon itself.

```
                          SORTER

        This program will sort up to 10 variables at once. Sorted
data will be printed on the screen in nicely formatted columns
if space permits.  Long labels will be truncated for printing
if necessary.  Files may contain up to 500 observations
(either numeric or alphabetic data).  Sorted data may be
saved in files if you wish.  Two options exist:

        1. Sort each variable separately
        2. Sort all variables on one variable

Which option (1 or 2)? 2
How many files (maximum is 10)? 4
Name of file 1 ? country
Name of file 2 ? b-rate
Name of file 3 ? gnp-cap
Name of file 4 ? density


DATA TYPE IN EACH FILE:

File COUNTRY  : Alphabetic Data
File B-RATE   : Numeric Data
File GNP-CAP  : Numeric Data
File DENSITY  : Numeric Data


OPTION TO SORT ALL VARIABLES ON ONE

Variable list:    1 = COUNTRY
                  2 = B-RATE
                  3 = GNP-CAP
                  4 = DENSITY

Sort on which variable number? 4


   Case   COUNTRY        B-RATE  GNP-CAP  DENSITY
   --------------------------------------------------
     1    Chad             44       220       10
     2    Bolivia          47       890       12
     3    Norway           13     14930       33
     4    Tanzania         47       230       54
     5    U.S.A.           15     11590       64
     6    Afghanistan      32       170       66
     7    Egypt            38       540      110
     8    Guatemala        43      1150      174
     9    Indonesia        43       460      192
    10    Spain            16      5600      200
    11    Vietnam          42       150      434
    12    Bangladesh       47       110     1665
   --------------------------------------------------


OPTION TO SAVE THE DATA

Want to save the sorted data (y/n)? n

Where now:  E=EXPLORE menu   R=Run SORTER again   X=exit ? e
```

■ **Exercises**

22.1 Using the DATABASE-1 diskette, run SORTER with two files: STATE (the state names) and UNEM (percent unemployment). Sort all the states on UNEM, being sure to list the file STATE first (so that the states will be properly identified). Which five states had the highest unemployment rates? The lowest? What factors might explain this pattern?

22.2 Choose two or three state data files from the DATABASE-1 diskette that you think might be related to the level of unemployment. Run SORTER using the files STATE, UNEM, and the two or three other files you selected. Sort all the variables on UNEM, being sure to list the file STATE first (so the states will be properly identified). Do the sorted UNEM values seem to correspond to the sorted values of the other two or three variables you chose? If so, what does it suggest? If not, what does it mean? Discuss fully.

23
SPLIT/Splitting a Data File

Introduction

From the moment that we first learn to summarize sample data (with a program such as ANALYZ or BOXPLOT) we begin to wonder what might be causing the observed pattern of variation. Why does Utah have a high birth rate and Connecticut a low one? Why does Alaska have a low cancer rate and Florida a high rate? Questions like this lead us toward *statistical inference* (as opposed to *descriptive statistics*).

Without sophisticated statistical tests we can use a "sample split" or "sample dichotomy" to make comparisons that can reveal much about our data. The idea is to divide a sample (variable Y) into two subgroups based on another variable (variable X) that we suspect might explain some of the variation in Y. For example, income levels among regions (high-income versus low-income regions) might help to explain infant mortality patterns. Or federal defense contracts (high-defense versus low-defense regions) might help explain variation in unemployment rates. The program SPLIT automates the task of sample-splitting by splitting a data file into two subfiles. Then we can use a program such as ANALYZ or BOXPLOT to compare the means, medians, standard deviations, and histogram shapes for our two subgroups (subfiles). More advanced programs (such as ANOVA or TWOSAM) can perform further tests.

One advantage of a sample split is that it gives you a built-in point of comparison. Another advantage of sample splits is that they let you do an original investigation—to find out something that nobody else knows or that is not in any textbook. The DATABASE-1 and DATABASE-2 files give you access to data with which to perform sample splits on your own.

Finally, sample-splitting helps you get ready for more advanced topics, such as t tests, F tests, and regression, which can help you answer questions such as "Are the observed differences between groups significant?"

■ Case Study 23.1: Income and Infant Mortality

To investigate whether infant mortality rates are related to per-capita income, we divide fifty states into two equal groups, depending on whether they are above or below the median income (based on the state database file IN-COME). Then we split infant mortality rates (state database file INFMOR) into two subfiles (call them INFMOR1 and INFMOR2) corresponding to the low-income and high-income states. These tasks are handled by program SPLIT (see the sample printout at the end of the chapter). SPLIT shows the sample dichotomy and allows us to print the two groups of data if desired. We save our two new subfiles on disk for analysis.

Next, we run ANALYZ (not shown) to compare our two saved subfiles INFMOR1 and INFMOR2. From ANALYZ we record the sample statistics and display them side by side, as shown in Table 23.1. Results should be heavily rounded (Ehrenberg, 1981) in order for the presentation to be effective.

In writing a synopsis we would note that the samples have similar shape (skewness, kurtosis). But high-income states have lower infant mortality rates (lower mean and lower quartiles) and exhibit less variation (smaller standard deviation and coefficient of variation).

By preparing two histograms with the same scale (using ANALYZ) we can see the situation at a glance. Figure 23.1 makes it clear that no high-income state has an infant mortality rate exceeding 16, while six of the low-income states do. Differences in modality and dispersion can also be seen in the histogram without doing any formal tests.

On the average, it is correct to say that poor states have higher infant mortality than rich states. But differences of this modest magnitude (14.0 versus 12.9) would be hard to detect in practice. In other words, despite the difference in means, some poorer states may have lower infant mortality rates

Table 23.1   Infant Mortality Rates per 1000 Births

	Low-Income States (n = 25)	High-Income States (n = 25)
Mean	14.0	12.9
S.D.	2.4	1.2
1st Quartile	12.2	11.9
2nd Quartile	14.1	13.0
3rd Quartile	15.8	13.8
Skewness	0.2	0.2
Kurtosis	2.3	2.4

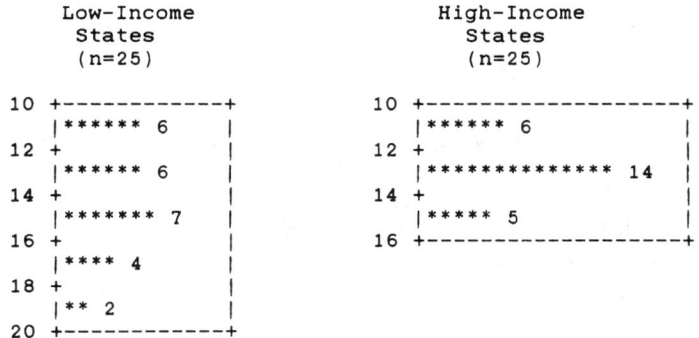

```
        Low-Income                        High-Income
          States                            States
          (n=25)                            (n=25)

   10 +------------+                 10 +-------------------+
      |****** 6    |                    |****** 6          |
   12 +            |                 12 +                  |
      |****** 6    |                    |************** 14 |
   14 +            |                 14 +                  |
      |******* 7   |                    |***** 5           |
   16 +            |                 16 +-------------------+
      |**** 4      |
   18 +            |
      |** 2        |
   20 +------------+
```

Figure 23.1   Histograms of Infant Mortality Rates

than some richer states. For example, Alaska has the highest income ($12,790) of any state, but its infant mortality rate (14.4) exceeds Utah's (11.4), which has a relatively low income ($7,649). In fact, variation *within* samples may dominate variation *between* samples. We would ask the general question "How large is the region of overlap in the histograms?" as illustrated stylistically in Figure 23.2.

Differences in means do not always imply readily observed differences in individual cases. The same is true in many other situations. For example, some seat belt users do not survive accidents, some basketball stars are undiscovered until college, and some geniuses become college professors despite poor schools. But in general, seat belts *do* save lives, basketball stars *are* recruited in high school, and geniuses *cannot* realize their intellectual potential with a poor education.

If we want to identify the states within each group, we can run SPLIT again, this time splitting the state names (file STATE) on per-capita income (file INCOME). The computer will print the names of the states in each group, listed in ascending order within each group (lowest to highest). When we split

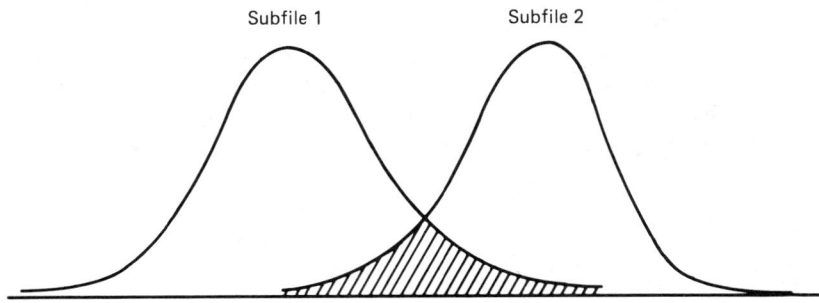

Figure 23.2   Variation within Samples versus between Samples

STATE on INCOME, the twenty-five low-income states (INCOME < 9348) will look like this:

MISS S C ARK ALAB KENT UTAH TENN W VA S D
N C VER N M ME IDA GA LA MONT N D ARIZ IND
MO FLA OKLA N H ORE

Within the low-income states, Mississippi is at the bottom and Oregon at the top (almost at the median). The twenty-five high-income states (INCOME $\geq$ 9348) will look like this:

WISC IOWA NEB VA PENN R I OHIO TEX MINN MICH
KANS COLO HAW MASS N Y WASH DEL MD ILL NEV
WYO N J CAL CONN ALAS

Wisconsin is just above the median, while Alaska has the highest income of any state. If you wish, you may save these names in two new files (say, STATE1 and STATE2) to use as labels in ANALYZ. But perhaps you want only to list them for later reference.

In effect, SPLIT mechanizes an operation that you could perform by hand if you listed X and Y observations as side-by-side columns, visually divided the cases into two groups using any dichotomy on X, and entered the corresponding Y values into two new files. The *median split* can be performed by sorting Y on X (using program SORTER) as illustrated in Table 23.2. Then you could save the sorted Y in a new file and use the EXPLORE file editor to split the saved file. SPLIT combines these steps efficiently, but it is easy to lose track of what is going on. After running SPLIT you may find it helpful to run SORTER to obtain a side-by-side listing of the two files of interest, to verify that the split was done the way you really wanted, and to serve as an appendix to reports you write.

You have several other options with SPLIT. First, you can experiment with cutoff points other than the median. For example, we could choose $10,000 as the income cutoff point (instead of $9,348). The sample sizes would have been unequal (36 low-income states, 14 high-income states). There is no reason that the sample sizes must be identical, though logic tends to lead us toward somewhat balanced splits.

You also have the option of splitting the sample on a binary variable. For example, we could have chosen WEST (1 for western states, 0 otherwise) to divide infant mortality rates into two groups (western versus nonwestern states). SPLIT can recognize a binary splitting file. You can also use the EX-PLORE editor to create your own binary variable (for example, ERA = 1 for states that approved the Equal Rights Amendment, ERA = 0 for the others) and use your new binary file to split other files (for example, to study the relationship between ERA approval and other variables). Binary variables are a very powerful tool, once you understand them.

Table 23.2   Infant Mortality Rates Sorted on Income

Low-Income States			High-Income States		
State	INCOME	INFMOR	State	INCOME	INFMOR
MISS	6580	18.7	WISC	9348	11.2
S C	7266	18.6	IOWA	9358	12.6
ARK	7268	16.4	NEB	9365	13.0
ALAB	7488	16.1	VA	9392	13.8
KENT	7613	12.7	PENN	9434	13.7
UTAH	7649	11.4	R I	9444	13.6
TENN	7720	14.8	OHIO	9462	13.3
W VA	7800	15.1	TEX	9545	14.3
S D	7806	13.5	MINN	9724	12.0
N C	7819	16.6	MICH	9950	13.8
VER	7827	10.4	KANS	9983	12.5
N M	7841	14.1	COLO	10025	11.2
ME	7925	10.4	HAW	10101	11.1
IDA	8056	11.7	MASS	10125	11.1
GA	8073	15.4	N Y	10260	14.0
LA	8458	17.3	WASH	10309	12.5
MONT	8536	11.6	DEL	10339	13.2
N D	8747	13.5	MD	10460	14.7
ARIZ	8791	13.1	ILL	10521	15.7
IND	8936	13.1	NEV	10727	12.5
MO	8982	14.8	WYO	10898	13.0
FLA	8996	14.1	N J	10924	13.0
OKLA	9116	14.3	CAL	10938	11.8
N H	9131	10.4	CONN	11720	11.6
ORE	9317	12.9	ALAS	12790	14.4

SPLIT

This program will split a file into two subfiles based on another file. For example, you could split a file DIVORCE (divorce rates in 50 states) into two subgroups by sorting on UNEM (unemployment rates for 50 states) to get:

```
DIVORCE1 = divorce rates in 25 low-unemployment states
DIVORCE2 = divorce rates in 25 high-unemployment states
```

You may then compare the subfiles (using ANALYZ or BOXPLOT) or test for significant differences (using ANOVA or TWOSAM).

```
The split will be at the median, unless the median of the
splitting file isn't unique.  You also have the option of
choosing your own cutoff point.  The program can recognize
the special case of a binary splitting file (such as WEST=0
or WEST=1).  The file to be split must be the same size as
the splitting file.  Any type of data may be in the file to
be split, but the splitting file must be numeric.

Name of file to be split? infmor
Name of file to split on? income
Use median as cutoff point (y/n)? y


INFMOR : 25 items in first group (INCOME < 9348 )

 18.7  18.6  16.4  16.1  12.7  11.4  14.8  15.1  13.5  16.6
 10.4  14.1  10.4  11.7  15.4  17.3  11.6  13.5  13.1  13.1
 14.8  14.1  14.3  10.4  12.9

INFMOR : 25 items in second group (INCOME >= 9348 )

 11.2  12.6  13.0  13.8  13.7  13.6  13.3  14.3  12.0  13.8
 12.5  11.2  11.1  11.1  14.0  12.5  13.2  14.7  15.7  12.5
 13.0  13.0  11.8  11.6  14.4

Please supply file names for the two subgroups of INFMOR.
If these files already exist, you may halt the save.  File
names should be 8 characters or less (e.g. INFMOR1 and INFMOR2).
If you don't want to save, press Enter without the file names.
Default drive will be assumed is you press the Enter key.

Which drive for saving the files (A, B, C, or D)? c
Save INFMOR for low-INCOME  group in which file? infmor1
Save INFMOR for high-INCOME group in which file? infmor2

File C:INFMOR1 successfully saved.
File C:INFMOR2 successfully saved.

Hint: Now you might want to run SORTER to obtain a nice listing
of INFMOR sorted on INCOME with a label file (e.g. state names).

Where now:   E=EXPLORE menu    R=Run SPLIT again    X=exit ? e
```

◼ Exercises

23.1 Use SPLIT to dichotomize TDEATH (traffic fatality rates per 100,000 persons in the fifty states of the United States) into two subfiles TDEATH1 and TDEATH2 using the potential explanatory variable UND25 (percent of drivers under age 25). The split will yield

Group 1:   States with few young drivers (file TDEATH1),

Group 2:   States with many young drivers (file TDEATH2).

Save your two new subfiles. Briefly, write down your expectations about how these two groups might differ in central tendency, and why. If you can think of more than one argument about cause-and-effect, write them both down. If alternative lines of reasoning lead you to contradictory expectations, note this fact.

23.2   Run ANALYZ (and/or BOXPLOT) twice to calculate a statistical profile of each subfile (TDEATH1 and TDEATH2). Be sure the histograms have comparable intervals (if not, rerun ANALYZ). Print the screens you need.

23.3   Run SPLIT again, this time splitting file STATE on UND25 so that you can obtain the names of the states in each group. Print the important screens. You may want to save the state names in subfiles so that you can use them as labels in ANALYZ (you could call them STATE1 and STATE2).

23.4   Write a simplified side-by-side table to compare the key statistics for these two subfiles, rounding heavily to ease the comparison. Then write a succinct summary of your conclusions about possible differences in central tendency, dispersion, and shape (skewness, kurtosis). Would you say that TDEATH1 and TDEATH2 are basically the same, or could the differences be significant? Is there a consistent pattern in the quartiles? If so, what does it imply?

23.5   Discuss the appearance of the histograms. Is there any apparent difference in central tendency? Do the two distributions appear to overlap a great deal? If the difference is slight, what conclusion do you draw?

23.6   Are there outliers? If so, which states are they? Is there a serious adverse effect on your analysis? What could be done about it?

23.7   Consider the states in each group. Can you suggest reasons for their relative positions? Where does your state fall in terms of traffic fatality rates? Does its position seem logical on the basis of what you know about other states?

23.8   Could other explanatory factors be at work, possibly obscuring the relationship? Name some omitted factors, and discuss their possible effects.

23.9   Repeat Exercises 23.1–23.8, using any two variables of your choice from the STATE database. Let your imagination be your guide. Just choose

Y = a variable whose pattern of variation is of interest (variable to be split),

X = a variable that is a proposed cause of variation in Y (variable to split on).

Optional Exercises for Advanced Learners

23.10   Repeat Exercises 23.1–23.8, using a different cutoff point than the median to see whether it makes any difference.

23.11   Repeat Exercises 23.1–23.8, creating your own data files from an almanac. For example, compare death rates from cirrhosis of the liver in states with high versus low consumption of alcohol per capita.

**Table 23.3 Performance of Randomly Selected
Growth-Oriented Mutual Funds**

Fund	Rating	Risk	Return	Fund	Rating	Risk	Return
ABT Utility Inc	C−	4	18.1	Mass Cap Dev	A+	4	42.9
Alliance Surveyor	D	4	28.4	Merrill Lynch Bas Val	B+	3	27.7
Am Cap Enterprise	C	3	26.9	Merrill Lynch Spc Val	C−	3	26.6
Am New Economy	C	3	21.1	Mutual Shares Corp	A+	2	27.8
BLC Growth	D	3	28.7	National Aviat & Tech	D	4	19.1
Bullock Canadian	D	3	4.4	NEL Growth	A−	4	33.5
Charter Fund	C	3	32.1	Nicholas Fund	A	3	25.5
Copley Tax Managed	A−	3	32.8	Oppenheimer Re- gency	C	3	17.9
Dean-Witter Ind Val	C	3	36.4	Penn Mutual	A	3	26.7
Dreyfus Third Cent	C+	3	24.7	Pro Fund	C−	3	32.3
Euro Pacific Growth	A+	3	40.7	Pru-Bache Research	A	3	43.8
Fidelity Discoverer	A	3	31.5	Putnam Investors	C−	4	24.4
Fidelity Magellan	A+	4	52.0	Reich & Tang Equity	A	3	39.8
Fidelity Selct—Dfns	C+	3	19.9	Scudder Cap Growth	A−	3	37.5
Fidelity Selct—Leis	A+	4	52.2	Seligman Comm & Info	B−	3	27.2
Fidelity Capital Gro	B−	3	32.3	Sequoia Fund	A−	2	25.9
Franklin Equity	C	3	30.8	Sigma Special	C	3	25.0
Fund of the Southwst	D	3	24.0	Sigma Venture Shares	C	4	27.2
GT Pacific	C+	3	33.7	Stein Roe Universe	C	3	35.5
IDS Growth	C	4	31.3	T Rowe Price New Era	C−	3	28.5
Investors Research	B	3	25.8	Templeton Growth	B+	3	27.2
JP Growth	C	3	24.0	Twentieth Cen Growth	C+	4	45.8
Keystone Intl	B−	3	48.2	United Accumulative	C	3	22.0
Lehman Capital	B+	3	27.5	USAA Cornerstone	A−	3	24.3
Lindner Fund	A+	2	19.9	Vanguard Sp Ptf - Hth	A+	3	45.5

Source: *Money,* Vol. 15, No. 5, May 1986, pp. 208–234.

23.12 Repeat Exercises 23.1–23.8, using two variables chosen from Table 23.3, which shows one-year rates of return for fifty growth-oriented mutual funds chosen at random from a population of 183 growth-oriented funds. For example, you could use the EXPLORE editor to enter the twelve-month percent return in a file called RETURN and then split this file on a second file named RISK (using a dichotomy like low-risk = 2 or 3, high-risk = 4 or 5). Or you could enter the overall performance in a file called RATING and use it to split RETURN (low-rating = D or C, high-rating = B or A). If you use the ratings, they must be converted to a numerical scale (such as A+ = 12, A = 10, A− = 9, and so on) before you do any calculations.

The sampling methodology was to select every fourth company on each page of the list. The rating is an assessment of the fund's overall performance over various time periods (not just one year). Risk is a measure of volatility, 1 representing the most stable and 5 the most volatile. The twelve-month return is in percent for the period ended April 1, 1986. (Caution: these data are intended solely for classroom purposes and should not be viewed as a guide to investments).

■ Reference

Ehrenberg, A. S. C., "The Problem of Numeracy," *The American Statistician,* Vol. 35, 1981, pp. 67–71.

24 TRANS/
Transformations of Data Files

Introduction

Statisticians often need to transform a variable by taking logs or absolute values or by raising it to a power. They also need to create new variables that are the sums, differences, or products of other variables. These transformations can be tedious if performed by hand but are quite simple if done by a computer program such as TRANS.

TRANS attempts to anticipate the most common needs of those learning to use statistics. Transformations will generally not be needed until you get into regression and should not be treated as beginners' tools. But they are available if needed.

TRANS may also be used to demonstrate certain statistical principles. For example, one of the standard theorems presented to learners is that if a random variable X has mean and standard deviation

$$\mu_X = \text{mean of X}$$
$$\sigma_X = \text{standard deviation of X}$$

then the linearly transformed variable Y defined as

$$Y = aX + b$$

will have a mean and standard deviation equal to

$$\mu_Y = a\mu_X + b$$
$$\sigma_Y = a\sigma_X.$$

Using TRANS, the validity of this theorem can easily be demonstrated, using any data file you choose. In particular, learners might find it interesting that adding a constant has no effect on the standard deviation. There is also a transformation to allow us to verify that the standardized variable

$$Z = \frac{X - \mu_X}{\sigma_X}$$

has a mean of zero and a standard deviation of one.

These are only examples of the types of exercises that can be created by using TRANS. A little imagination will suggest others. For example, the generalized difference transform (explained in Chapter 6's "Technical Notes") is widely used to adjust for autocorrelation in time-series data, using

$$X'_t = X_t - \rho X_{t-1} \quad (t = 2, 3, \ldots, n)$$

with the special transformation

$$X'_1 = X_1 \sqrt{1 - \rho^2}$$

for the first observation. You may estimate ρ by using the autocorrelation estimate from BIGRES or MGRES, or you may input $\rho = 1$ to obtain ordinary first differences. The case studies illustrate other practical uses for advanced statistical research.

Case Study 24.1: Squaring a Variable

Suppose we want to create a new regression predictor EDUC2, to be equal to the square of the variable EDUC. The program TRANS can be used to carry out this transformation. Table 24.1 shows how the file arbitrarily called EDUC2 would look after being created from the twenty original data values

Table 24.1   Creating a Squared File

Employee	EDUC	EDUC2
Mary	6	36
Frieda	6	36
Alicia	4	16
Tom	4	16
Nicole	2	4
Bill	0	0
Gillian	6	36
Bob	2	4
Vivian	4	16
Cecil	0	0
Barney	2	4
Jack	0	0
Wanda	6	36
Sam	2	4
Saundra	4	16
Pete	4	16
Steve	6	36
Fred	2	4
Dick	2	4
Lee	2	4

of the predictor EDUC. Each value of EDUC2 is the square of the corresponding value of EDUC.

This example refers to twenty employees whose college education (in years) is of interest in a regression experiment (see MGRES Case Study 15.1 for further details). The new predictor EDUC2 could be used to test a nonlinear model such as

$$\text{SALARY} = \beta_0 + \beta_1 \text{ EDUC} + \beta_2 \text{ EDUC2} + \text{Random Error}$$

as opposed to just using a linear model such as

$$\text{SALARY} = \beta_0 + \beta_1 \text{ EDUC} + \text{Random Error.}$$

If MGRES shows that extra nonlinear term has a coefficient that is significantly different from zero (indicated by a large t value), we may conclude that the relationship between SALARY and EDUC is quadratic (parabolic) instead of linear.

```
                              TRANS

        This program offers transformations of one or two
     variables.  For example, you might need to square X to get
     a new variable X-SQR for a non-linear regression.  Or you
     might need an interaction variable AGE-SAL, which is the
     product of AGE and SALARY.  Or you might divide CRIME by POP
     to obtain a new variable CRIMRATE, the crime rate per capita.
     The new variable is printed on a new file.  You will be
     warned if the file is not empty.  Maximum is N = 1000 cases.

     --------------------------------------------------------------
     |   Transformations of    |   Transformations of           |
     |   One Variable X:        |   Two Variables X and Y:       |
     |-------------------------|--------------------------------|
     |     1. X**a             |      10. X*Y                    |
     |     2. a*X + b           |      11. X/Y                    |
     |     3. (X-Mean)/SD       |      12. X+Y                    |
     |     4. abs(X)            |      13. X-Y                    |
     |     5. log(X)            |      14. a*X + b*Y              |
     |     6. ln(X)             |      15. a*X/Y                  |
     |     7. 10**X             |                                |
     |     8. e**X              |                                |
     |     9. X - r * X         |                                |
     |         t        t-1     |                                |
     |-------------------------|--------------------------------|
     |        [Note: a and b are any constants]                  |
     --------------------------------------------------------------
     Input number of the transformation you want? 1
     Input the constant a : 2
     Which file contains X? educ


     PROCEDURE TO SAVE TRANSFORMED DATA

     Your new variable X**a has been created.
     Which drive for saving the file (A, B, C, or D)? c
     Name of file to receive the new variable? educ2

     File C:EDUC2 successfully saved.

     Where now:  E=EXPLORE menu   R=Run TRANS again   X=exit ? e
```

■ **Case Study 24.2: Creating an Interaction Term**

Suppose the statistician would like to know whether two variables AGE and EDUC exhibit an interactive effect in a regression model, which might not be accounted for by either variable alone. This could be expressed in a regression model such as

$$\text{SALARY} = \beta_0 + \beta_1 \text{ AGE} + \beta_2 \text{ EDUC} + \beta_3 \text{ AGE\&EDUC} + \text{Random Error.}$$

The new predictor (arbitrarily called AGE&EDUC) is the product of the two variables AGE and EDUC. The printout shows how TRANS can be used to create this new predictor, and Table 24.2 shows what the variables would look like after the transformation.

If MGRES reveals that the estimated coefficient of the new interaction term AGE&EDUC is significantly different from zero (as evidenced by a large t value), the statistician could conclude that the interaction between AGE and EDUC is important. This type of experiment is sometimes important, espe-

Table 24.2   Creating an Interaction Term

Employee	AGE	EDUC	AGE&EDUC
Mary	20	6	120
Frieda	31	6	186
Alicia	44	4	176
Tom	20	4	80
Nicole	55	2	110
Bill	44	0	0
Gillian	25	6	150
Bob	55	2	110
Vivian	50	4	200
Cecil	60	0	0
Barney	40	2	80
Jack	64	0	0
Wanda	35	6	210
Sam	64	2	128
Saundra	40	4	160
Pete	31	4	124
Steve	25	6	150
Fred	35	2	70
Dick	60	2	120
Lee	50	2	100

cially in behavioral research. Of course, there are other ways of studying interaction effects (for example, two-way ANOVA).

```
                         TRANS

      This program offers transformations of one or two
variables.  For example, you might need to square X to get
a new variable X-SQR for a non-linear regression.  Or you
might need an interaction variable AGE-SAL, which is the
product of AGE and SALARY.  Or you might divide CRIME by POP
to obtain a new variable CRIMRATE, the crime rate per capita.
The new variable is printed on a new file.  You will be
warned if the file is not empty.  Maximum is N = 1000 cases.

 ---------------------------------------------------------
| Transformations of        | Transformations of         |
| One Variable X:           | Two Variables X and Y:     |
|---------------------------|----------------------------|
|    1. X**a                |    10. X*Y                 |
|    2. a*X + b             |    11. X/Y                 |
|    3. (X-Mean)/SD         |    12. X+Y                 |
|    4. abs(X)              |    13. X-Y                 |
|    5. log(X)              |    14. a*X + b*Y           |
|    6. ln(X)               |    15. a*X/Y               |
|    7. 10**X               |                            |
|    8. e**X                |                            |
|    9. X - r * X           |                            |
|       t        t-1        |                            |
|---------------------------------------------------------|
|         [Note: a and b are any constants]               |
 ---------------------------------------------------------

Input number of the transformation you want? 10
Which file contains X? age
Which file contains Y? educ


PROCEDURE TO SAVE TRANSFORMED DATA

Your new variable X*Y has been created.
Which drive for saving the file (A, B, C, or D)? c
Name of file to receive the new variable? age&educ

File C:AGE&EDUC successfully saved.

Where now:  E=EXPLORE menu   R=Run TRANS again   X=exit ? e
```

Exercises

24.1 Use option 2 of TRANS to create a variable NEWDOCS, obtained by transforming the variable DOCS on the DATABASE-1 diskette:

$$NEWDOCS = 3 * DOCS + 100.$$

24.2 Run ANALYZ on DOCS and NEWDOCS. Compare the mean of DOCS and the mean of NEWDOCS. Compare the standard deviations. What would you expect? What is the effect of this linear transformation on the other statistics, such as skewness and kurtosis? Discuss fully.

24.3 Use option 3 of TRANS to create a variable ZDOCS, using DOCS as the input file. Run ANALYZ on ZDOCS to verify that the standardized file has zero mean and unit standard deviation.

Optional Exercises for Ambitious Learners

24.4 Use option 2 of TRANS to create a new data file named DOCS2 by squaring the state data file DOCS on the DATABASE-1 diskette. Then run MGRES using INFMOR as the dependent variable and both DOCS and DOCS2 as independent variables:

$$INFMOR = f(DOCS, DOCS2).$$

Is there evidence of a nonlinear relationship between INFMOR and DOCS? How would a nonlinear relationship look on a graph of INFMOR against DOCS?

24.5 Use option 1 of TRANS to create a new time-series predictor YEAR2, the square of YEAR on the DATABASE-1 diskette. Then use MGRES to estimate the quadratic trend relationship:

$$GNP = f(YEAR, YEAR2).$$

Discuss what the results tell you about linear trend and quadratic trend. How would such a relationship appear on a graph of GNP against YEAR? How does this model compare with GNP trends that would be estimated by TREND?

24.6 Use option 5 of TRANS to create a new variable LOGGNP, to be the logarithm of the time-series variable GNP on the DATABASE-1 diskette. Then use BIGRES to estimate the trend relationship:

$$LOGGNP = f(YEAR).$$

How does the fit compare with the simple trend regression model:

$$GNP = f(YEAR).$$

How do both models compare with the quadratic trend model you estimated in Exercise 24.5? Discuss fully.

24.7 Run BIGRES to estimate a simple savings function PS = f(PI) using the time-series database files. Use the estimated autocorrelation coefficient from BIGRES with TRANS option 9 to perform a generalized difference transform, producing new files NEWPS and NEWPI. Then rerun BIGRES using the model NEWPS = f(NEWPI). Discuss your findings.

25

TREND/Fitting Trends to Time-Series Data

Introduction

In a rapidly changing world we usually cannot make effective decisions merely by looking at the present. We also need to know where we have been and where we are going. Consequently, decision makers in business, labor, and public administration spend a great deal of time assembling and interpreting time-series data.

A time series is any numerical quantity measured at regular time intervals. Most publicly distributed time-series data are available annually (crime rates), quarterly (GNP), or monthly (unemployment rate). But we may be interested in data collected as often as every day (Dow-Jones stock price index) or only once per decade (U.S. population census). Modern businesses maintain an impressive array of internal time-series data bases, reflecting their sales, profits, costs, production, inventory, and so on.

The program TREND will extract the overall *trend* of a nonseasonal time series. Annual data are typically used, though other data are acceptable if seasonally adjusted. The trend gives an overview of the pattern of growth (or decline) with detailed fluctuations averaged out. Trends help reveal the "big picture," even though they are simplifications. Using the Ordinary Least Squares method, TREND fits three models:

Linear: $Y_t = A + Bt$
Exponential: $Y_t = AB^t$
Quadratic: $Y_t = A + Bt + Ct^2$

where Y_t = value of Y in period t (t = 1,2, . . . ,n) and A, B, and C are constants to be determined by TREND.

TREND displays fluctuations about the trend and uses the trend to project forecasts into the future. Various statistics of fit are calculated for each model. TREND also plots a graph to give a quick visual impression of the data.

When trends are projected into the future, we are implicitly assuming that the future will be like the past. Forecasters call such models *naive* because they ignore the underlying causes of the change in Y. Yet naive forecasting models are widely used because they are objective and easy to understand. In the short run, naive forecasts can be quite successful, since underlying conditions do persist over considerable periods of time (for example, sales keep rising, prices keep rising). Of course, when underlying conditions change, this kind of fore- cast will miss the mark entirely. Trend forecasts are frequently modified by experience and intuition (broadly classified as *judgment inputs*). At a mini- mum, trends give a baseline on which to build.

■ Case Study 25.1: Growth of Federal Debt

In the *Wall Street Journal* we often read stories about the possible economic effects of federal government borrowing (when tax collections fail to keep pace with federal spending). We know that the federal debt keeps rising be- cause of such borrowing. But just how large is the federal debt, and how fast is it changing? To put things in perspective, we can calculate the federal gov- ernment debt *per person,* by dividing total current-dollar federal debt by total U.S. population, yielding the time series shown in Table 25.1.

Apparently, the average American's share of the federal debt is nearing $8000. This may not seem like much, since many Americans borrow twice that much just to finance a new car (not to mention a house). Some experts worry, however, because the debt is growing rapidly. We can see at a glance that federal debt per capita has more than doubled in less than a decade.

To obtain a more precise idea of the change in federal debt per capita, we use the EXPLORE editor to create a data file named FEDDEBT containing these nine data points. Then we run TREND to estimate the various trend

Table 25.1   U.S. Federal Government Debt per Capita, 1977–1985

Year	Per Capita Federal Debt
1977	3219
1978	3505
1979	3704
1980	4015
1981	4363
1982	4936
1983	5891
1984	6660
1985	7651

Source: *Economic Report of the President* (Washington, D.C.: U.S. Government Printing Office, 1986), pp. 283, 339.

models (see the printout below). We define 1977 as the base year so that the computer can label the years correctly. We ask for three years of forecasts, plus a graph.

```
                         TREND

        This program fits several trend models to a time
series, assesses the fit, and makes future forecasts.
Using trends, you can identify periods when your observed
data were unusually high or low.  Program expects a data
file consisting of a single column of numbers representing
your actual time series values over the past N periods
(maximum N = 200).  Three trend models will be fitted:

               1. Exponential trend
               2. Linear trend
               3. Quadratic trend

You will be shown statistics to assess fit.  A graph
will be printed if your data set isn't too big.



QUESTIONS AND OPTIONS

Which file contains your data? feddebt
May I use graphics characters (y/n)? y
Table of actual and estimated FEDDEBT values (y/n)? y
Are you using yearly data (y/n)? y
Enter initial year (e.g. 1967)? 1977
Want any forecasts (y/n)? y
How many periods ahead? 3
Want a graph (y/n)? y
Input short capitalized graph caption (e.g. WIDGET SALES)
or just press ENTER if file name FEDDEBT is good enough?
```

TABLE OF ACTUAL AND ESTIMATED FEDDEBT VALUES

Time	Actual	Exponential Trend	% Error	Linear Trend	% Error	Quadratic Trend	% Error
1977	3219	3031.90	5.8	2716.80	15.6	3316.07	-3.0
1978	3505	3380.13	3.6	3258.27	7.0	3408.08	2.8
1979	3704	3768.36	-1.7	3799.73	-2.6	3628.51	2.0
1980	4015	4201.18	-4.6	4341.20	-8.1	3977.36	0.9
1981	4363	4683.70	-7.4	4882.67	-11.9	4454.61	-2.1
1982	4936	5221.65	-5.8	5424.13	-9.9	5060.29	-2.5
1983	5891	5821.39	1.2	5965.60	-1.3	5794.38	1.6
1984	6660	6489.99	2.6	6507.07	2.3	6656.88	0.0
1985	7651	7235.40	5.4	7048.53	7.9	7647.81	0.0
1986		8066.43		7590.00		8767.14	
1987		8992.90		8131.47		10014.90	
1988		10025.79		8672.93		11391.06	

The 1988 forecasts vary widely. The linear model projects the smallest 1988 per capita debt (8673). The quadratic forecast is the largest (11391), while the exponential forecast lies in between (10026). This difference in forecasts suggests that it *does* matter which model we use.

Looking at the table of actual and forecast Y values, we see that the linear and exponential models seem too low in 1977 and too high in 1980–82. By 1985, actual debt is beginning to outrun the linear forecasts. The linear model clearly has larger errors than the other two, which begins to cast doubt on its validity for making forecasts. The forecast of the quadratic model, in contrast, is almost perfect in 1984–85.

```
           SUMMARY OF FIT FOR THE THREE MODELS OF FEDDEBT
=================================================================
                    Exponential      Linear        Quadratic
                       Model          Model           Model
-----------------------------------------------------------------
Mean Absolute
Per Cent Error         4.2            7.4             1.7

Standard
Error               261.9473        435.7099        99.27041

R-Squared            0.9734          0.9298          0.9969


-----------------------------------------------------------------
```

From the summary-of-fit table we see that the quadratic and exponential models have a fairly small mean absolute percent error (MAPE), calculated by taking the average of the percent errors with their signs removed:

Linear: MAPE = 7.4 percent;

Exponential: MAPE = 4.2 percent;

Quadratic: MAPE = 1.7 percent.

The linear model has a noticeably larger MAPE than the other two models, which again shows that a nonlinear growth model is preferable.

Another statistical measure of fit is r^2, which lies between 0 and 1. If the actual and estimated Y values were equal (that is, if all the trend predictions were perfect), then r^2 would be exactly equal to 1. Conversely, when r^2 is near 0, we conclude that there is no meaningful trend. For federal debt, TREND's calculations show that r^2 is quite near its maximum for all three models:

Linear: $r^2 = .9298$

Exponential: $r^2 = .9734$

Quadratic: $r^2 = .9969.$

It is customary to say that r^2 measures the "percent of variation in Y explained by trend." For example, the quadratic trend explains 99.7 percent of the variation in federal debt per capita over time. By this criterion the quadratic model gives the best fit, though not by much. On the basis of r^2 alone, it is actually rather difficult to choose between the two best models.

Because the quadratic model has more constants (A, B, and C), it will often give a better fit than the other two models (the linear is a special case of the quadratic, with C = 0, so the quadratic always gives at least as high an r^2 as the linear model). But is the added complexity worthwhile?

The English philosopher William of Occam (1300–1349) proposed that, confronted with two equivalent explanations, we should prefer the simpler. This is the principle of *Occam's Razor*. The linear and exponential models are appealing because of their simplicity. Therefore we would prefer the quadratic model only if it gave *substantially* better results than the other two. For FEDDEBT, we will reserve judgment until we have examined other measures of fit.

Another measure of fit is the standard error (denoted SE), which is based on the sum of the squared prediction errors for our sample:

$$SE = \sqrt{\frac{\Sigma[Y_t - \hat{Y}_t]^2}{n - m}}.$$

Here n is the sample size (number of periods), and m is the number of constants estimated. The standard error is measured in the same units as Y (such as dollars). If actual and estimated Y values are equal in all time periods (an unlikely event), SE will be zero. The larger the SE, the worse the fit. For the linear and exponential models, m = 2 (because we estimate A and B); for the quadratic model, m = 3 (because we estimate A, B, and C). Rounding things off a bit, the respective standard errors calculated by TREND are

Linear: SE = 436;

Exponential: SE = 262;

Quadratic: SE = 99.

The quadratic model clearly gives the best fit. You might find it surprising that the standard errors differ so much, given the similarity of the r^2 for our three models. Forecasters pay close attention to the standard error because it often gives a better idea of the *practical utility* of a model and because it is a more sensitive barometer of fit than some other measures.

Using the standard error, we can construct an approximate 95 percent confidence interval for our forecast by adding or subtracting twice the standard error. For the quadratic model our 1988 forecast would be

$$\hat{Y} \pm 2SE$$

$$\hat{Y} \pm 2(99)$$

$$11391 \pm 198.$$

Thus our rough 95 percent prediction limits would range from Y = 11,193 to Y = 11,589. If we wanted to forecast only the mean of Y, it would be approximately correct to divide SE in the preceding formula by the square root of the sample size (n = 9), which would obviously narrow the interval considerably:

$$\hat{Y} \pm 2SE/\sqrt{n}$$
$$\hat{Y} \pm 2(99)/\sqrt{9}$$
$$11{,}391 \pm 66.$$

Our rough 95 percent mean prediction limits for 1988 would range from Y = 11,325 to Y = 11,457.

```
ESTIMATED TREND MODELS OF FEDDEBT

1. Exponential Trend Model   [for t=1,2, ... , 9 ]
                                       t
   Y =    2719.55 (   1.11486)

   Compound growth rate =   11.5 per cent
   Exponential trend is very significant (t-statistic = 16.007)


2. Linear Trend Model        [for t=1,2, ... , 9 ]

   Y =    2175.33    +    541.467 t

   Linear trend is very significant (t-statistic = 9.626)


3. Quadratic Trend Model     [for t=1,2, ... , 9 ]
                                                  2
   Y =    3352.48    -    100.611 t   +   64.2078 t

   Linear trend is marginally significant (t-statistic =  -1.735)
   Quadratic trend is very significant (t-statistic = 11.351)
```

Now consider the fitted trend equations, rounded off a bit. For each model the starting point in the initial period (t = 0) is the intercept (denoted by A or Y_0):

$$\text{Exponential:} \quad \hat{Y}_t = 2720(1.115)^t$$
$$\text{Linear:} \quad \hat{Y}_t = 2175 + 541t$$
$$\text{Quadratic:} \quad \hat{Y}_t = 3352 - 101t + 64t^2.$$

The intercepts are similar in the exponential (A = 2720) and linear (A = 2175) models, while the quadratic model's intercept (A = 3352) is larger.

Even though its fit is in doubt, the linear model has one virtue: Its slope is easy to interpret. The linear model says that federal debt per person is growing at an average rate of about $541 per year. The linear model assumes that Y changes by a constant amount each period, though this assumption may not be realistic.

Growth in the exponential model is nonlinear:

$$Y_t = AB^t.$$

If $B > 1$, then Y will increase in each period by an increasing absolute amount (but a constant percentage rate). If $B < 1$, then Y will decrease in each period by a decreasing absolute amount (but a constant percentage rate). We may see the meaning of the exponential model more clearly by writing

$$B = 1 + r$$

where r represents the compound rate of growth (if $r > 0$) or decline (if $r < 0$). Since the intercept is $A = Y_0$, the exponential model may be written

$$Y_t = Y_0(1 + r)^t$$

which most readers will recognize as the standard formula for compound interest with Y_0 as the beginning balance.

For the debt data,

$$\hat{Y}_t = 2720(1.115)^t$$

giving

$$B = 1 + r$$

or

$$B = 1 + .115.$$

or $r = .115$ or 11.5 percent per annum. This implied compound growth rate can be compared with growth rates from other TREND runs (for example, we could estimate the rate of growth of GNP) or with events from our own experience (such as rates of return on our own investments). It would be hard to convince most taxpayers that their incomes are rising at an equivalent rate per annum!

The other nonlinear model is the quadratic model, which shows a negative, marginally significant linear component and a positive, significant quadratic component:

$$Y_t = A + Bt + Ct^2$$
$$\hat{Y}_t = 3352 - 101t + 64t^2.$$

If the trend continues, an even more rapid rate of change will exist in the future, since the quadratic component will begin to dominate the equation (positive second derivative).

Unlike the other two models, the quadratic model can have a turning point (that is, a peak or trough), which may be found by calculus. The signs of B and C determine the shape of the quadratic trend, as illustrated in Figure 25.1. Estimated federal debt per capita has passed its turning point (because $B < 0$ and $C > 0$).

One final technical point concerns the meaning of the phrase "significantly different from zero." TREND finds the standard deviation associated with each estimated constant in the trend equation and calculates the ratio of

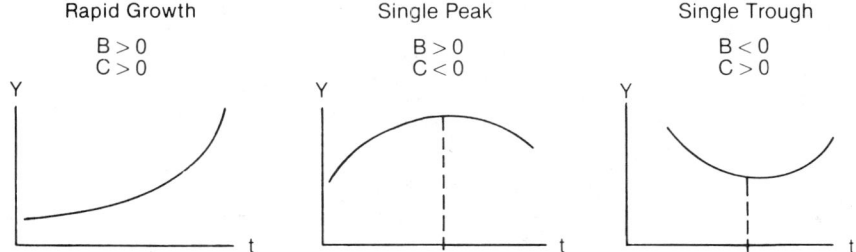

Figure 25.1   Archetypal Quadratic Trend Shapes

the estimated constant to its standard deviation. If this ratio (called a t statistic) is large in absolute magnitude, it suggests that the estimated coefficient is significantly distant from zero. The rules used by TREND for determining significance are

| t statistic | $\geq$ 3 $\rightarrow$ Very significant trend;

| t statistic | $\geq$ 2 $\rightarrow$ significant trend;

| t statistic | $\geq$ 1 $\rightarrow$ marginally significant trend;

| t statistic | < 1 $\rightarrow$ trend not significant.

These rules of thumb will work reasonably well as long as the sample size is not too small.

The sample printout shows that *all* of the trend terms in *all* of the models are highly significant except the linear term in the quadratic model. This tends to confirm the fit we have observed using the various other measures such as MAPE, r², and SE.

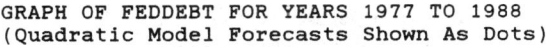

GRAPH OF FEDDEBT FOR YEARS 1977 TO 1988
(Quadratic Model Forecasts Shown As Dots)

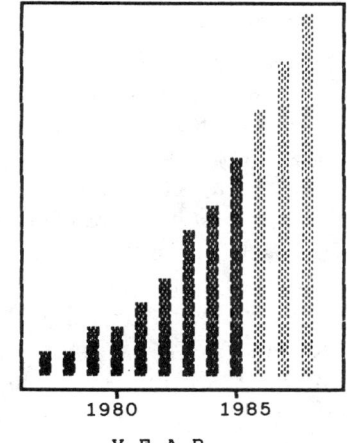

Where now: E=EXPLORE menu   R=Run TREND again   X=exit ? e

Absolute Error	Exponential	Linear	Quadratic
0.0–1.9%	xx	x	xxxx
2.0–3.9%	xx	xx	xxxxx
4.0–5.9%	xxxx		
6.0–7.9%		xx	
8.0% or more		xxxx	

Figure 25.2 Tabulation of Errors for Three Models

The graph printed by TREND reveals why the linear model fits poorly. By holding a straightedge to the graph and sketching a linear trend line we can see that the linear model fits poorly on both ends (too low) and in the middle (too high).

As a final check, we can make a frequency tabulation of error magnitudes for all three models, as shown in Figure 25.2. On the basis of this tabulation it is apparent that the quadratic model's percentage errors are most frequently close to the mark, with the exponential model close behind. The linear model's errors are much larger.

The quadratic model has the best fit. But are its forecasts reliable? Are nine years of data enough? Probably so, because the data cover a relevant period of time, and the progression in Y is fairly smooth. There is no compelling reason to include pre-1977 data. The period 1977–1985 covers a cross-section of economic events: periods of rapid inflation, periods of recession, Republican and Democratic administrations, and so on.

To cover their bets, forecasters sometimes present three forecasts: lower bound, best guess, and upper bound. This approach is helpful because it leaves the final choice to the reader of the forecaster's report.

The real question is, Will the federal deficit per capita continue to grow, as forecast? Couldn't the future differ from the past? Of course, a tremendous economic boom, large government spending cuts, or a tax increase could reduce federal deficits. But it seems more likely that the root causes of deficits will continue: inflation, recession, tax cuts (which are politically popular), and federal government spending increases (which are popular with both military and civilian lobbyists). Given the past behavior of political leaders, the causes of this statistical trend seem likely to remain.

To be believable, a forecast needs at least a verbal underpinning of qualifications and assumptions, which we have tried to provide here in narrative form. But to have long-run predictive power, more complex models are necessary. We should not make excessive claims about our quadratic trend model, even though it seems satisfactory for short-run forecasting.

One final question: When does statistics give way to economics? Establishing a significant statistical trend tells us what *is* happening but not necessarily what *should* happen. Should we be alarmed about this trend, which has now persisted for many years? Or will our nation get along somehow, more or less as we have done so far?

Some experts feel that the government ought to try to balance its budget, at least when the economy is booming, for fear that federal borrowing might crowd out private borrowers, whose needs are thought to be more legitimate than those of the government. Part of the problem has been that the economy has not been booming during much of the past decade. But shouldn't the government still try to live within its means, come good or bad times?

Other experts might suggest that the government cannot be expected to behave more responsibly than individuals or firms, who regularly amass new debts for purposes that are not always productive or worthwhile. They would point to rapid growth rates in consumer debt, business debt, corporate salaries, average hourly wages and fringes, and so on. Some experts believe that it is better to look at federal debt as a fraction of GNP (a measure of national income) for the same reason that banks use individual income to establish individual borrowing limits. Finally, many experts feel that deficits are appropriate when the economy is sluggish.

Further statistical tasks await, no matter what position one takes on the deficit. Comparing trends is a time-honored statistical technique that will continue to be used by participants on both sides of the debate about federal deficits and the national debt.

■ Exercises

25.1 Choose one of the time-series data files already contained on the DATABASE-1 diskette. Run TREND, making forecasts for what you think is a reasonable number of periods ahead. Make a printed copy of the results from the screen, if possible. If not, rerun the programs as often as needed to answer Exercises 25.2–25.9.

25.2 Study the estimated trend equations. Compare the intercepts. Try to interpret (in real-world meaning) each term in the three models, especially the slope of the linear model and average percent growth rate in the exponential model. Discuss the shape of the quadratic model, based on the signs of its coefficients. Which terms in these three models are significant, and which are not? What is the implication for your model choice?

25.3 Discuss the pattern of errors over time in each of the three models. Do the models fit as well at the beginning as at the end? What about the middle years? Make a frequency tabulation of errors, and discuss which model seems to perform most consistently. Discuss fully.

25.4 Discuss the summary measures of fit, including mean absolute percentage error (MAPE), standard error (SE), and r^2. Can you rule out any model on grounds that it is clearly a worse fit than the others? Does any one model clearly emerge as having the best fit, or are the models similar?

25.5 On the basis of all factors (significance of coefficients, overall fit, error patterns, error frequency, Occam's Razor), which model do you prefer? Discuss fully.

25.6 Discuss the underlying causes of the observed trend (or lack of trend if the fit is poor). Can you think of possible explanations for the periods when the model(s) fit poorly? Is there a pattern in the errors? If so, can you explain it?

25.7 Do you believe the forecasts? Why, or why not?

25.8 Examine the graph. Describe its appearance. Try drawing a nicely labeled graph of your own. Sketch TREND's three forecasts on your hand-drawn graph (using different-colored pens). Do your forecasts seem reasonable?

25.9 Try writing a concise executive synopsis of your findings, suitable for quick assimilation by a busy decision maker. Highlight your main findings only, and suppress unnecessary detail. What would be gained by reading your report, without ever seeing the raw data? Would anything important be lost?

Optional Exercises for Ambitious Students

25.10 Construct an approximate 95 percent confidence interval for your best model's forecasts. How wide is the interval, expressed as a percent of the forecast? Discuss.

25.11 Use the EXPLORE file editor to create a diskette data file containing one of the three time-series data sets shown in Tables 25.2–25.4. Be careful to

Table 25.2 Appeals Filed in U.S. Appeals Court		Table 25.3 U.S. Sheep and Lamb Population		Table 25.4 Hospital Cost per Patient Day	
Year	Cases	Year	Beasts (mil.)	Year	Dollars
1971	12,788	1967	23.9	1970	74
1972	14,535	1968	22.1	1971	83
1973	15,629	1969	21.2	1972	95
1974	16,436	1970	17.4	1973	102
1975	16,658	1971	19.7	1974	113
1976	18,408	1972	18.7	1975	133
1977	19,118	1973	17.7	1976	152
1978	18,918	1974	16.4	1977	173
1979	20,219	1975	14.5	1978	194
1980	23,200	1976	13.3	1979	216
1981	26,362	1977	12.7	1980	244

Source: U.S. Bureau of the Census, *Statistical Abstract of the United States, 1982–83* (Washington, D.C.: U.S. Government Printing Office, 1983).

use file names that do not conflict with files already stored on the diskette. For example, you could use APPEALS and SHEEP and HOSPCOST for these three respective time-series variables. Then repeat Exercises 25.2–25.9. If possible, check your forecasts against actual data from the library.

25.12 Use a home or library data source such as the *Statistical Abstract of the United States* or the *Economic Report of the President* to find a time series that interests you. Use the EXPLORE data file editor to create a diskette data file. Be careful to avoid using file names that might conflict with existing data files. Then repeat Exercises 25.2–25.9.

TWOSAM/Two-Sample Comparisons

Introduction

TWOSAM permits three kinds of hypothesis tests, probably the most commonly used in basic statistics: test of two means, test of two variances, and test of two proportions.

The t test for two sample means is probably the most useful of all everyday hypothesis tests because so many real-life situations involve comparing Group A with Group B. For example, a women's organization might wish to compare average annual earnings for men and women workers in a certain profession. Personnel consultants might wish to compare average sick days claimed for employees in two office locations. Health economists might wish to compare the average cost of a certain operation in two different hospitals. Depending on your situation, we would choose one of the following pairs of hypotheses:

$$H_0: \quad \mu_1 = \mu_2 \qquad H_0: \quad \mu_1 = \mu_2 \qquad H_0: \quad \mu_1 = \mu_2$$
$$H_1: \quad \mu_1 < \mu_2 \qquad H_1: \quad \mu_1 \neq \mu_2 \qquad H_1: \quad \mu_1 > \mu_2$$

TWOSAM will carry out the t test for two means, either for two data files or using two already calculated sample means and standard deviations. The calculation is first carried out assuming equal population variances, then again assuming unequal population variances (with Welch's Behrens-Fisher correction for degrees of freedom). You can later decide which version to use.

The F test for equality of two sample variances is closely related to the t test. It is treated by TWOSAM as part of the test for two sample means, to allow you to decide which assumption is most reasonable (equal or unequal variances).

The test for significant difference between two sample proportions is natural for describing many experimental situations. Does Group A have a higher proportion of persons with cavities than Group B (testing toothpaste)? Does absenteeism in Plant A differ significantly from that in Plant B (trying a worker incentive plan)? Does the proportion of defective computer chips produced by the day shift differ from that produced by the night shift (setting quality control standards)? In each case we are comparing the hypothesis that two population proportions are equal against the alternative hypothesis that the two population proportions differ. Possible pairs of hypotheses are:

$$H_0: \quad p_1 = p_2 \qquad H_0: \quad p_1 = p_2 \qquad H_0: \quad p_1 = p_2$$
$$H_1: \quad p_1 < p_2 \qquad H_1: \quad p_1 \neq p_2 \qquad H_1: \quad p_1 > p_2$$

An automatic check for sample size is performed, and TWOSAM prints a warning if the assumption of normality is in doubt.

◼ Case Study 26.1: Surgical Costs at Two Hospitals

Auditors for the Jolly Blue Giant health care insurance company are studying the cost of appendectomies performed at Carver Memorial Hospital and Hackmore General Hospital. Samples from recent patient records reveal the statistics shown in Table 26.1. These data suggest that Carver's average cost is lower, but could this be simply due to sample variation? The hypotheses are

$$H_0: \quad \mu_1 = \mu_2 \qquad \text{(null hypothesis).}$$
$$H_1: \quad \mu_1 < \mu_2 \qquad \text{(suggested by sample).}$$

No data files are needed for this problem. We just enter the sample results when we run TWOSAM to obtain the printout on page 270. Let's review the calculations performed by TWOSAM.

A glance at the two standard deviations suggests that equal population variances can be assumed (because s_1 and s_2 are similar in size). Using the formula for equal population variances, TWOSAM's estimated standard deviation of the difference of the two sample means is

$$\hat{\sigma}_{\bar{x}_1 - \bar{x}_2} = \sqrt{\frac{(n_1 - 1)s_1^2 + (n_2 - 1)s_2^2}{n_1 + n_2 - 2}} \quad \sqrt{\frac{1}{n_1} + \frac{1}{n_2}}$$

$$= \sqrt{\frac{(21 - 1)(107)^2 + (13 - 1)(126)^2}{21 + 13 - 2}} \quad \sqrt{\frac{1}{21} + \frac{1}{13}}$$

$$= (114.49508)(.352905)$$

$$= 40.405919.$$

Table 26.1   Cost of Inpatient Appendectomies

Carver Hospital	Hackmore Hospital
$\bar{X}_1 = \$2,100$	$\bar{X}_2 = \$2,220$
$s_1 = \$107$	$s_2 = \$126$
$n_1 = 21$ patients	$n_2 = 13$ patients

Note that the first part of this formula (TWOSAM's "pooled" standard deviation of about 114.50) lies between the two sample standard deviations, as would be expected ($s_1 = 107$ and $s_2 = 126$). If we put the standard deviation of the difference of means into the formula for the test statistic, we can verify TWOSAM's calculation:

$$t = \frac{\overline{X}_1 - \overline{X}_2}{\hat{\sigma}_{\overline{x}_1 - \overline{x}_2}} = \frac{2100 - 2220}{40.405919} = -2.970.$$

The critical value of t from Appendix C will require d.f. = 32, which TWOSAM obtains by combining the degrees of freedom for both samples:

$$\text{d.f.} = (n_1 - 1) + (n_2 - 1) = (21 - 1) + (13 - 1) = 32.$$

If we choose the .10 level of significance, the left-tail decision rule will look like Figure 26.1.

Since the test statistic ($t = -2.970$) is far to the left of the critical value ($t = -1.310$), we will reject H_0 in favor of H_1. We conclude that Carver Hospital really does have lower mean cost for simple inpatient appendectomies.

At the .10 level of significance it is not really very difficult to reject H_0. However, our decision would also hold up at more stringent levels of significance. For example, at the .01 level of significance (with d.f. = 32) we have a higher critical value ($t = -2.457$), but we can still reject H_0.

So the observed difference in means is quite significant. Whether you think $120 is enough to matter might depend on whether you are paying the bill from your own pocket or you have health care insurance. But either way, it is going to come from somewhere, and to most people it would be a matter of practical importance, as well as of statistical significance.

Now suppose we had assumed unequal variances. Then the standard deviation of the difference of means becomes

$$\hat{\sigma}_{\overline{x}_1 - \overline{x}_2} = \sqrt{\frac{s_1^2}{n_1} + \frac{s_2^2}{n_2}} = \sqrt{\frac{107^2}{21} + \frac{126^2}{13}} = 42.02881.$$

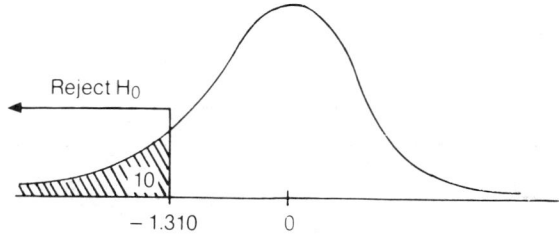

Reject H_0

.10

−1.310   0

Figure 26.1   t Test for Two Sample Means

Therefore TWOSAM's test statistic is

$$t = \frac{\overline{X}_1 - \overline{X}_2}{\hat{\sigma}_{\overline{x}_1 - \overline{x}_2}} = \frac{2100 - 2220}{42.02881} = -2.855.$$

The "bottom line" is that the assumption we make about variances does not affect the test statistic very much ($t = -2.855$ versus $t = -2.970$) because the numerator is exactly the same and the denominator is approximately the same (42.03 versus 40.41).

But when unequal variances exist, we also ought to use a correction for the Behrens-Fisher problem in degrees of freedom. With the Welch's Behrens-Fisher correction the resulting degrees of freedom will be less than the sum of the separate sample degrees of freedom. Using Welch's Behrens-Fisher correction, TWOSAM gets

$$d.f. = \frac{[s_1^2/n_1 + s_2^2/n_2]^2}{\dfrac{[s_1^2/n_1]^2}{n_1 - 1} + \dfrac{[s_2^2/n_2]^2}{n_2 - 1}}$$

$$= \frac{[107^2/21 + 126^2/13]^2}{\dfrac{[107^2/21]^2}{21 - 1} + \dfrac{[126^2/13]^2}{13 - 1}} = 22.4$$

TWOSAM rounds off (conservatively) to the next lower degrees of freedom to get d.f. = 22. But the decision is the same as before because the critical t value from Appendix C changes only slightly ($t = -1.321$ versus $t = -1.310$). So even the rather complicated-looking Welch's Behrens-Fisher correction does not affect our decision.

TWOSAM does these calculations both ways (using equal and unequal variances). You can look at the results and decide for yourself how much difference the assumption of equal variances makes, what effect the Behrens-Fisher correction has on degrees of freedom, and so on. This is an illustration of what computers can do best. Freed from mind-fogging work, you can do what people do best: Exercise judgment.

Either assumption about variances will tend to yield about the same test statistic, as long as the two sample sizes and sample standard deviations are not too dissimilar. Even Welch's Behrens-Fisher correction often will not make much difference in the critical t value unless the sample sizes are tiny. Without a computer, the Welch's Behrens-Fisher correction is a lot of work. One possible way around it is to use the unequal variance formula with d.f. = n − 1, where n is the smaller of the two sample sizes. This is a conservative simplifying assumption that makes the work easier.

What does it matter whether or not we assume equal variances? Why not always assume unequal variances, for simplicity, since there always will be slight differences between the sample standard deviations? In fact, this is sometimes done to save time. But you could lose power in the test.

All preceding calculations may be verified by looking at the sample print-out of TWOSAM that follows. Obviously, it is a lot easier to obtain the results from the printout, rather than from the manual calculations illustrated here.

```
                        TWOSAM

This program will test for significant differences between:

        1. Two sample means
        2. Two sample proportions

Which option do you want (1 or 2)? 1

TEST FOR DIFFERENCE OF TWO SAMPLE MEANS

This program will compute Student's t to test for significant
difference between two sample means.  It will also compute
the F statistic to test for equality of two sample variances.
You may input your two sample means and standard deviations
directly, or you may read two data files.  Each data file
should contain one column of data.  The two sample sizes
(n1 and n2) may be unequal.  Maximum is 1000 items per file.

Are you using stored data files (y/n)? n

For Sample 1: Input Mean? 2100
              Input S.D.? 107
              Input N   ? 21

For Sample 2: Input Mean? 2220
              Input S.D.? 126
              Input N   ? 13


TABLE OF SUMMARY FACTS

Sample 1        Sample 2
--------        -------
  2100            2220        Sample Mean
   107             126        Sample Standard Deviation
 11449           15876        Sample Variance
    21              13        Sample Size

Note: The above formulas for variances and
      standard deviations divide by n-1

TEST FOR DIFFERENCES BETWEEN TWO SAMPLES:

1. Test of Two Means: Assuming Equal Population Variances

        Student's t =   -2.970  with d.f.=  32

2. Test of Two Means: Assuming Unequal Population Variances

        Student's t = -2.855  with d.f.=  32 ( d.f. =  22 using
                                  Welch's Behrens-Fisher correction)

3. Test For Equality of Population Variances (Inverted Ratio)

        F =     1.387 with d.f.= 12 for numerator
                      and d.f.= 20 for denominator

4. Pooled Estimate of Population Variance

        Estimated Pooled Variance = 13109.13
        Estimated Pooled Std. Dev.= 114.4951

Where now:  E=EXPLORE menu   R=Run TWOSAM again   X=exit ? e
```

■ Case Study 26.2: Variance in Surgical Costs

When is it really safe to assume equal variances? There is a simple test that can easily be made using a calculator (or using the program TWOSAM). Suppose we want to test these two hypotheses about the two population variances:

$$H_0: \quad \sigma_1^2 = \sigma_2^2.$$

$$H_1: \quad \sigma_1^2 > \sigma_2^2.$$

The test statistic is the ratio of the sample variances, which should not be far from unity (F = 1) if H_0 is true:

$$F = s_1^2/s_2^2 = \frac{\text{larger estimated variance}}{\text{smaller estimated variance}}.$$

To ensure a right-tail test, we will put the larger variance in the numerator. The degrees of freedom for the F ratio will correspond to the numerator and denominator degrees of freedom for the two sample variances:

Numerator: d.f. $= n_1 - 1$

Denominator: d.f. $= n_2 - 1$.

For convenience we just say that F has d.f. $= (n_1 - 1, n_2 - 1)$. If we have to reverse the sample variances to ensure an F ratio greater than unity, we just interchange the numerator and denominator degrees of freedom. This procedure may be illustrated by using our two hospital samples in Table 26.1.

It appears that Hackmore has a larger variance than Carver. But could the observed difference in standard deviations be due to sample variation? After all, the sample sizes are fairly small. To find out, we will calculate the F ratio using the squared sample standard deviations. This calculation is illustrated here (see the preceding TWOSAM printout for details). TWOSAM's right-tail test statistic is larger than unity, suggesting at least some departure from H_0:

$$F = (126)^2/(107)^2 = 1.39.$$

To ensure a right-tail test, TWOSAM places the second sample variance in the numerator (on the screen, TWOSAM will print "Inverted Ratio" to let you know). Therefore TWOSAM must also switch the numerator and denominator degrees of freedom:

Numerator: d.f. $= 13 - 1 = 12$

Denominator: d.f. $= 21 - 1 = 20$.

Using Appendix E, we obtain a critical value (F = 2.35) using the .05 level of significance for d.f. $= (10,20)$. The decision rule is shown in Figure 26.2. Note that we use the next lower degrees of freedom if the desired value cannot be found in Appendix E—a conservative procedure that will not increase the chance of Type I error. We cannot reject the hypothesis of equal variances, since our test statistic (F = 1.39) does not exceed the critical value (F = 2.35).

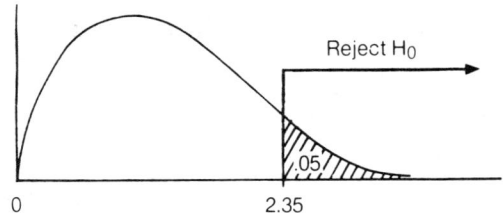

Figure 26.2   F Test for Equality of Variances

If you look at various F values in Appendix E, you will see that generally one variance must be several times as large as the other for you to conclude that unequal variances exist. Thinking back to the t test, this means that except in extreme cases, you will find yourself using an equal population variance t test when you do a test for the difference of two means.

The F test result (for equal/unequal population variances) tells you which formula to use when you do the t test for difference of two means. Therefore the F test and the t test are related. If you choose your t test (equal or unequal variance formula) by taking a glance at s_1 and s_2, you are performing a casual F test by the eyeball method, which may be sufficient in some cases.

In theory, the F test for equal/unequal variances should precede the t test. In practice, if you use a computer program like TWOSAM, you will use the F test to decide which t test is appropriate after the fact. The computer does it both ways at no extra cost. By looking at the F ratio from TWOSAM you can subsequently decide which result to use.

For our hospital example the implication is that the equal-variance t test is preferred over the unequal-variance t test:

$$\text{Assuming equal variances:} \quad t = -2.970.$$
$$\text{Assuming unequal variances:} \quad t = -2.855.$$

In this example it does not matter which test is used, since we get the same decision anyway. This is often the case unless the sample sizes and variances differ greatly.

■ Case Study 26.3: American League versus National League

Is there a statistically significant difference between averages of batting champions in the American and National Leagues? Consider the data for the period 1950–1981, shown in Table 26.2.

In the previous example we had already calculated the sample means and standard deviations, using a pocket calculator. But in this baseball example we will enter the raw data and let the computer do all the math. Using the EXPLORE editor, we create a separate data file for each league. We call our

Table 26.2   Batting Averages for League Champions, 1950–1981

Year	American League Player and Club	Average	Year	National League Player and Club	Average
1950	Goodman (Bost)	.354	1950	Musial (StL)	.346
1951	Fain (Phil)	.344	1951	Musial (StL}	.355
1952	Fain (Phil)	.327	1952	Musial (StL)	.336
1953	Vernon (Wash)	.337	1953	Furillo (Brook)	.344
1954	Avila (Clev)	.341	1954	Mays (NY)	.345
1955	Kaline (Det)	.340	1955	Ashburn (Phil)	.338
1956	Mantle (NY)	.353	1956	Aaron (Milw)	.328
1957	Williams (Bost)	.388	1957	Musial (StL)	.351
1958	Williams (Bost)	.328	1958	Ashburn (Phil)	.350
1959	Kuenn (Det)	.353	1959	Aaron (Milw)	.355
1960	Runnels (Bost)	.320	1960	Groat (Pitt)	.325
1961	Cash (Det)	.361	1961	Clemente (Pitt)	.351
1962	Runnels (Bost)	.326	1962	Davis (LA)	.346
1963	Yastrzemski (Bost)	.321	1963	Davis (LA)	.326
1964	Oliva (Minn)	.323	1964	Clemente (Pitt)	.339
1965	Oliva (Minn)	.321	1965	Clemente (Pitt)	.329
1966	Robinson (Balt)	.316	1966	Alou (Pitt)	.342
1967	Yastrzemski (Bost)	.326	1967	Clemente (Pitt)	.357
1968	Yastrzemski (Bost)	.301	1968	Rose (Cinn)	.335
1969	Carew (Minn)	.332	1969	Rose (Cinn)	.348
1970	Johnson (Cal)	.329	1970	Carty (Atlan)	.366
1971	Oliva (Minn)	.337	1971	Torre (StL)	.363
1972	Carew (Minn)	.318	1972	Williams (Chi)	.333
1973	Carew (Minn)	.350	1973	Rose (Cinn)	.338
1974	Carew (Minn)	.364	1974	Garr (Atlan)	.353
1975	Carew (Minn)	.359	1975	Madlock (Chi)	.354
1976	Brett (KC)	.333	1976	Madlock (Chi)	.339
1977	Carew (Minn)	.388	1977	Parker (Pitt)	.338
1978	Carew (Minn)	.333	1978	Parker (Pitt)	.334
1979	Lynn (Bost)	.333	1979	Hernandez (StL)	.344
1980	Brett (KC)	.390	1980	Buckner (Chi)	.324
1981	Lansford (Bost)	.336	1981	Madlock (Pitt)	.341

Source: *Hammond Almanac, 1982* (Maplewood, N.J.: Hammond Almanac, Inc., 1982), pp. 934–935.

two data files AMER and NATL. Each file contains thirty-two observations (from 1950 to 1981 in Table 26.2).

We choose a two-tail hypothesis test because there is no prior indication which way the difference will lie (although sports fans may have their own ideas). Our hypotheses are

$$H_0: \quad \mu_{AMER} = \mu_{NATL}.$$

$$H_1: \quad \mu_{AMER} \neq \mu_{NATL}.$$

From the TWOSAM printout we can see that the sample means look about the same. Although the American League mean (.340) is slightly lower than the National League mean (.343), the difference is not impressive, so we do not expect to reject H_0.

In this problem, it does not matter whether or not equal population variances are assumed, owing to the identical sample sizes (the test statistic will be the same using either formula). However, we can see that the American League appears to be less consistent, judging by its larger standard deviation (.0214 versus .0109 for the National League).

Before we look at the t tests we should at least ask whether the observed inequality of sample variances is significant. With d.f. = (31,31) at the .05 level of significance we find a critical value (F = 1.84) from Appendix E. (We use the next lower degrees of freedom because the desired value is not in Appendix E—a conservative procedure that will not increase the probability of Type I error.) The F test decision rule is not illustrated because it is of secondary interest. The F test statistic from the TWOSAM printout (F = 3.879) is far in excess of the critical value (1.84), so we reject the hypothesis of equal population variances—the American League champions are in fact less consistent than the champions of the National League. Therefore we would prefer the t test using unequal variances.

For the main hypothesis test of two sample means we compare TWOSAM's test statistic from the printout (t = −.670) with a critical value from Appendix C (t = 2.014) for a two-tail decision rule at the .05 level of significance. The decision rule is illustrated in Figure 26.3.

In this test we use unequal variances, which according to the printout give us 45 degrees of freedom (using the Behrens-Fisher correction). The Behrens-

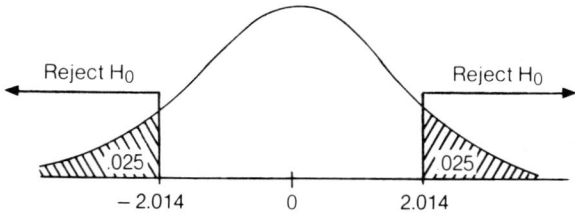

Figure 26.3   t Test for Two Means

```
                        TWOSAM

This program will test for significant differences between:

     1. Two sample means
     2. Two sample proportions

Which option do you want (1 or 2)? 1



TEST FOR DIFFERENCE OF TWO SAMPLE MEANS

This program will compute Student's t to test for significant
difference between two sample means.  It will also compute
the F statistic to test for equality of two sample variances.
You may input your two sample means and standard deviations
directly, or you may read two data files.  Each data file
should contain one column of data.  The two sample sizes
(n1 and n2) may be unequal.  Maximum is 1000 items per file.

Are you using stored data files (y/n)? y

Name of file #1? amer
Name of file #2? natl


   TABLE OF SUMMARY FACTS

   Sample 1       Sample 2
   --------       --------
    AMER           NATL        Data File Name
   .3400626       .3429062     Sample Mean
   .0214023       .0108668     Sample Standard Deviation
   .0004581       .0001181     Sample Variance
    32             32          Sample Size

   Note: The above formulas for variances and
         standard deviations divide by n-1



TEST FOR DIFFERENCES BETWEEN TWO SAMPLES: AMER VS. NATL

1. Test of Two Means: Assuming Equal Population Variances

        Student's t =   -0.670  with d.f.=  62

2. Test of Two Means: Assuming Unequal Population Variances

        Student's t =   -0.670  with d.f.=  62 ( d.f. = 45 using
                                    Welch's Behrens-Fisher correction)

3. Test For Equality of Population Variances

        F =     3.879 with d.f.= 31 for numerator
                      and d.f.= 31 for denominator

4. Pooled Estimate of Population Variance

        Estimated Pooled Variance = .0002881
        Estimated Pooled Std. Dev.= .0169727

Where now:  E=EXPLORE menu   R=Run TWOSAM again   X=exit ? e
```

Fisher correction really has little effect on the critical t value, since the combined sample size is fairly large. The decision is not even close. The difference in means is totally insignificant. On the average, batting champions in the two leagues are equally good.

Yet the feeling persists that the National League averages "look" better than those of the American League. Examining the individual data points, we notice that in twenty-two of the thirty-two years the National League champion has had a higher batting average than the American League champion. Suppose we take the difference between the two champions in each year, as shown in Table 26.3. This shows that the American League batter came out ahead in only ten out of thirty-two years. One American League player, Rod Carew of Minnesota, was personally responsible for five of the ten years in which the American League came out ahead. The other three American League standouts were Mickey Mantle of New York, Ted Williams of Boston, and George Brett of Kansas City. These four account for almost all of the instances in which the American League came out ahead.

Suppose we try a t test for *related samples,* using our column of n = 32 differences. This would actually be a more powerful t test because it makes use of the relatedness of the two samples. The mean of the column of differences is $\overline{d} = 0.002844375$, with a standard deviation of $s_d = .0227725$ (calculations by hand are not shown). The test statistic for the difference of two *related means* would be

$$t = \frac{\overline{d}}{s_d/\sqrt{n}} = \frac{.002844375}{.0227725/\sqrt{32}} = -.706.$$

We *still* conclude that there is no significant difference between the champions' batting averages in the two leagues, since the test statistic (t = −.706) does not even come close to the critical value (t = 1.697). Actual degrees of freedom would be d.f. = n − 1 = 31, but we use the next lower value in Appendix C because the desired value is not given.

Table 26.3   Difference between League Champions

Year	American Champion	National Champion	Difference
1950	.354	.346	.008
1951	.344	.355	−.011
.	.	.	.
.	.	.	.
.	.	.	.
1981	.336	.341	−.005

Case Study 26.4: Employee Job Satisfaction

The CBK Corporation administers a salary satisfaction questionnaire to randomly chosen employees in two of its departments, obtaining the statistics shown in Table 26.4.

The proportion of accountants who are satisfied with their jobs is about 64.0 percent, compared with only 53.7 percent for the programmer/analysts. Is this observed difference in proportions statistically significant, or could it be due to sample variation? We will use a right-tail test. The hypotheses are

$$H_0: \quad p_1 = p_2 \qquad \text{(equal population proportions)}.$$

$$H_1: \quad p_1 > p_2 \qquad \text{(suggested by sample)}.$$

If H_0 is true, both samples can logically be combined to estimate the single, common population proportion.

TWOSAM handles all the calculations (see the sample printout). The "pooled sample" proportion of satisfied employees is 60 percent, since we have 84 satisfied employees out of 140:

$$\bar{p} = \frac{X_1 + X_2}{n_1 + n_2} = \frac{55 + 29}{86 + 54} = \frac{84}{140} = .6.$$

If H_0 is true, then *both* samples came from a common population (60 percent satisfied), and sample variation only makes it *seem* that the two sample proportions differ (64 percent versus 54 percent). Assuming that the observed sample difference $\hat{p}_1 - \hat{p}_2$ is normally distributed, we may use the test statistic

$$z = \frac{\hat{p}_1 - \hat{p}_2}{\hat{\sigma}_{\hat{p}_1 - \hat{p}_2}} \ .$$

Assuming that H_0 is true, z should be near zero. However, owing to sample variation, $\hat{p}_1$ and $\hat{p}_2$ might differ, causing z to depart from zero. If z is *significantly* above zero, we will reject H_0. The standard deviation of the difference $\hat{p}_1 - \hat{p}_2$ is estimated by TWOSAM as follows:

$$\hat{\sigma}_{\hat{p}_1 - \hat{p}_2} = \sqrt{\frac{\bar{p}(1 - \bar{p})}{n_1} + \frac{\bar{p}(1 - \bar{p})}{n_2}}$$

Table 26.4   Satisfaction Rates among Employees

Accountants	Programmer/Analysts
$X_1 = 55$ generally satisfied	$X_2 = 29$ generally satisfied
$n_1 = 86$ respondents	$n_2 = 54$ respondents
$\hat{p}_1 = \dfrac{55}{86} = .6395$ or 64.0%	$\hat{p}_2 = \dfrac{29}{54} = .5370$ or 53.7%

$$\hat{\sigma}_{\hat{p}_1 - \hat{p}_2} = \sqrt{\frac{.6(1 - .6)}{86} + \frac{.6(1 - .6)}{54}} = .08506.$$

TWOSAM's test statistic is

$$z = \frac{\hat{p}_1 - \hat{p}_2}{\hat{\sigma}_{\hat{p}_1 - \hat{p}_2}} = \frac{.6395 - .5370}{.08506} = 1.205.$$

At the .05 level of significance our decision rule requires a critical value of 1.645 (from Appendix B), as shown in Figure 26.4. We cannot reject H_0, since the test statistic ($z = 1.205$) does not exceed the critical value ($z = 1.645$). The observed difference in sample proportions is within the range attributable to random chance.

In this example, altering the level of significance will not make much difference because $z = 1.205$ is really not an impressive departure from $z = 0$. The observed difference in sample proportions is not going to convince us to reject H_0 at any reasonable level of significance.

This is not to say that accountants and programmer/analysts definitely have equal salary satisfaction. We have not *proved* H_0, but have merely failed to *disprove* H_0. If our sample had been larger, the observed difference might have been significant. If we are sufficiently interested in the question to pursue it, we might state that the observed difference, "although not significant, is suggestive of greater salary dissatisfaction among programmer/analysts." We could recommend a follow-up sampling experiment at a later date.

We also could have chosen a two-tail test. In this problem a two-tail test would be reasonable if we have no *a priori* idea of which group will have greater salary satisfaction. Purists might prefer the two-tail test. Use your judgment.

It is a good idea to check that the sample size is large enough that normality may safely be assumed. In this case we use the rule of thumb that for each sample both np and $n(1 - p)$ should be at least 5 to justify the normality assumption. Since

$$n_1 \hat{p}_1 = (86)(.6395) = 55$$
$$n_2 \hat{p}_2 = (54)(.5370) = 29$$

and

$$n_1(1 - \hat{p}_1) = (86)(.3605) = 31$$
$$n_2(1 - \hat{p}_2) = (54)(.4630) = 25$$

we can safely assume normality. TWOSAM will carry out this calculation automatically and will print a warning if there might be a problem with the normality assumption.

What should you do if normality cannot be assumed? If the sample size is *almost* large enough, you might just go ahead and assume normality anyway (this is only a rule of thumb, not a law of science). Otherwise, things get a

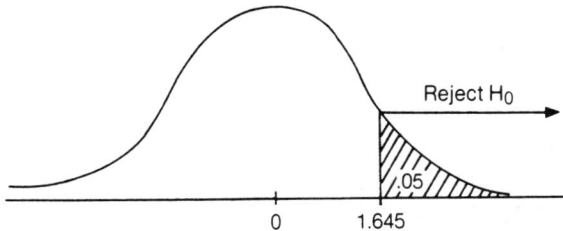

Figure 26.4   Test of Two Proportions Assuming Normality

little complicated because you will have to use two binomial distributions! In this situation you should invoke a failsafe emergency procedure:

$$< < < \text{"Call a professional statistician."} > > >$$

```
                        TWOSAM

This program will test for significant differences between:

        1. Two sample means
        2. Two sample proportions

Which option do you want (1 or 2)? 2


TEST FOR DIFFERENCE OF TWO SAMPLE PROPORTIONS

Each sample proportion must be a fraction within the
range 0 < p < 1.  Sample sizes will be requested.

First sample proportion:  p1 =.6395
First sample size:        n1 = 86

Second sample proportion: p2 =.5370
Second sample size:       n2 = 54

Calculated Test Statistic: Z = 1.205
           Pooled Proportion: p = .6000


Where now:  E=EXPLORE menu   R=Run TWOSAM again   X=exit ? e
```

■ Exercises

Instructions: Formulate the most reasonable pair of hypotheses about two proportions, two means, or two variances. Show all work clearly, including all formulas. Use the program TWOSAM for your calculations. Experiment with alternative levels of significance to see how sensitive the decision rule is.

26.1   Surgeons at Carver Memorial Hospital have a new procedure that might reduce the time required for an appendectomy. A sample of eleven appendectomies using the old method reveals a mean of thirty-eight minutes with a standard deviation of four minutes, while a sample of sixteen appendectomies with the new method reveals a mean of thirty-one minutes with a standard deviation of nine minutes. At the .10 level of significance, is the new procedure significantly faster? What would be the real-world consequence of Type I error? Type II? Are the variances equal? What difference would it make?

26.2   During a ten-week test period a control group of sixteen hogs had a mean weight gain of 140 pounds with a standard deviation of 15 pounds. A second group of twenty-six hogs received an experimental diet, showing a weight gain of 152 pounds with a standard deviation of 20 pounds. At the .01 level of significance, is the new diet effective in promoting weight gain? What would be the real-world consequence of Type I error? Type II? Are the variances equal? What difference would it make?

26.3   Two nationally known aviation training schools are being considered for the honor of instructing the Crown Prince of Petrolia. To see which school has a higher success rate on the FAA written exams, the Petrolian Ministry of Statistics surveyed graduates of both schools. According to the survey, 70 out of 140 graduates of Fly-Mor Academy passed their exams on the first try, compared with 104 out of 260 graduates of the Blue Yonder Institute. At the .05 level of significance, is there a significant difference between the success rates of the two schools? What would Type I error mean? Type II? May normality be assumed?

26.4   A random sample of fifty day-shift workers at Ramrod Manufacturing Corporation revealed that twenty-eight favored changing to flex-hours. A second random sample of fifty evening-shift workers revealed that twenty-two favored changing to flex-hours. At the .10 level of significance, is this a significant difference? What would be the real-world consequence of Type I error? Type II? May normality be assumed?

26.5   In the 1980 election, many analysts assumed that states with high unemployment would tend to favor the Democratic candidate, based on the historical political alignments in election years. The actual election results are shown in Table 26.5. Is this observed difference of means sufficient to convince you that the analysts were right? Discuss fully, using various levels of significance to see how sensitive the decision is. Use TWOSAM for all your calculations.

26.6   Is alcohol consumption significantly higher in states with high death rates due to cirrhosis of the liver? Table 26.6 indicates that a difference exists that can be tested using TWOSAM. Discuss your conclusion fully, using various levels of significance.

Table 26.5   1980 Unemployment Rates

States in Which Reagan Got at Least Half the Vote	States in Which Reagan Got Less Than Half the Vote
$\overline{X}_1 = 6.180$ percent	$\overline{X}_2 = 7.396$ percent
$s_1 = 1.560$ percent	$s_2 = 1.497$ percent
$n_1 = 25$ states	$n_2 = 25$ states

Source: G. J. Heil and K. Lhota, unpublished paper (Rochester, Mich.: Oakland University, 1983).

Table 26.6   1980 Alcohol Consumption per Capita

States with High Death Rate Due to Cirrhosis of the Liver	States with Low Death Rate Due to Cirrhosis of the Liver
$\overline{X}_1 = 13.824$ gallons	$\overline{X}_2 = 10.640$ gallons
$s_1 = 3.441$ gallons	$s_2 = 2.278$ gallons
$n_1 = 25$ states	$n_2 = 25$ states

Source: W. Hazelton, J. House, and A. Marchesi, unpublished paper (Rochester, Mich.: Oakland University, 1983).

Optional Exercises for Ambitious Learners

26.7   An experimental surgical procedure is being studied at Carver Hospital. Both procedures are considered safe. Five surgeons are asked to perform the same operation the old way and then on another patient using the new way. All patients are carefully matched by age, sex, and other relevant factors to ensure a valid comparison. The times required to perform the procedure, in minutes, are shown in Table 26.7. At the .05 level of significance, is the new way faster than the old way? What would be the real-world consequence of Type I error? Type II? In your analysis, do the test first as a related t test, then as independent samples. Which method is more appropriate, and why?

Table 26.7   Minutes Required to Perform Surgery

Surgeon	Old Way	New Way
1	36	31
2	55	45
3	28	28
4	40	35
5	62	57

26.8 Choose two samples of your own from any source so that you can compare the means. Set up hypotheses about the two means, choose a level of significance, and carry out the test. Use the EXPLORE editor to create two diskette data files. Some possible comparisons are

a. Traffic fatality rates in northern versus southern states of the United States (use your own definitions and file TDEATH from DATABASE-1 diskette).

b. Population density in eastern versus western states of the United States (use your own definitions and file POPDEN from DATABASE-1 diskette).

c. Unemployment rates in northern versus southern states of the United States (use your own definitions and file UNEM from DATABASE-1 diskette).

26.9 Choose one hundred words at random from this book. Find the proportion that have three or more syllables. Then choose one hundred words at random from a novel and do the same. Do a test to see if there is a significant difference between the two samples in the proportion with long words (three or more syllables). Choose your own level of significance and explain all your assumptions, including your methods of sampling.

26.10 Choose twenty words at random from this book and write down how many syllables each has. Do the same for a novel of your choice. For each sample, find the mean and standard deviation. Then do a test for significant difference of means. Choose your own level of significance. Explain any assumptions you made, including your sampling methods.

■ References

1. Brownlee, K. A., *Statistical Theory and Methodology in Science and Engineering* (New York: John Wiley and Sons, Inc., 1965), pp. 299–303.

2. Madsden, Richard W. and Moeschberger, Melvin L., *Statistical Concepts with Applications to Business and Economics* (Englewood Cliffs, N.J.: Prentice-Hall, 1980), p. 253.

3. Wang, Y. Y., "Probabilities of the Type I Errors of the Welch Tests for the Behrens-Fisher Problem," *Journal of the American Statistical Association*, Vol. 66, no. 335, pp. 605–608.

Appendixes

Appendix A: Normal Areas from 0 to Z

This table shows the shaded area under the curve.

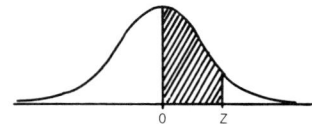

Z	Area	Z	Area	Z	Area	Z	Area	Z	Area	Z	Area
0.00	.00000	0.50	.19146	1.00	.34134	1.50	.43319	2.00	.47725	2.50	.49379
0.01	.00399	0.51	.19497	1.01	.34375	1.51	.43448	2.01	.47778	2.51	.49396
0.02	.00798	0.52	.19847	1.02	.34613	1.52	.43574	2.02	.47831	2.52	.49413
0.03	.01197	0.53	.20194	1.03	.34849	1.53	.43699	2.03	.47882	2.53	.49430
0.04	.01595	0.54	.20540	1.04	.35083	1.54	.43822	2.04	.47932	2.54	.49446
0.05	.01994	0.55	.20884	1.05	.35314	1.55	.43943	2.05	.47982	2.55	.49461
0.06	.02392	0.56	.21226	1.06	.35543	1.56	.44062	2.06	.48030	2.56	.49477
0.07	.02790	0.57	.21566	1.07	.35769	1.57	.44179	2.07	.48077	2.57	.49492
0.08	.03188	0.58	.21904	1.08	.35993	1.58	.44295	2.08	.48124	2.58	.49506
0.09	.03586	0.59	.22240	1.09	.36214	1.59	.44408	2.09	.48169	2.59	.49520
0.10	.03983	0.60	.22575	1.10	.36433	1.60	.44520	2.10	.48213	2.60	.49534
0.11	.04380	0.61	.22907	1.11	.36650	1.61	.44630	2.11	.48257	2.61	.49547
0.12	.04776	0.62	.23237	1.12	.36864	1.62	.44738	2.12	.48300	2.62	.49560
0.13	.05172	0.63	.23565	1.13	.37076	1.63	.44845	2.13	.48341	2.63	.49573
0.14	.05567	0.64	.23891	1.14	.37286	1.64	.44950	2.14	.48382	2.64	.49585
0.15	.05962	0.65	.24215	1.15	.37493	1.65	.45053	2.15	.48422	2.65	.49597
0.16	.06356	0.66	.24537	1.16	.37698	1.66	.45154	2.16	.48461	2.66	.49609
0.17	.06749	0.67	.24857	1.17	.37900	1.67	.45254	2.17	.48500	2.67	.49621
0.18	.07142	0.68	.25175	1.18	.38100	1.68	.45352	2.18	.48537	2.68	.49632
0.19	.07535	0.69	.25490	1.19	.38298	1.69	.45449	2.19	.48574	2.69	.49643
0.20	.07926	0.70	.25804	1.20	.38493	1.70	.45543	2.20	.48610	2.70	.49653
0.21	.08317	0.71	.26115	1.21	.38686	1.71	.45637	2.21	.48645	2.71	.49663
0.22	.08706	0.72	.26424	1.22	.38877	1.72	.45728	2.22	.48679	2.72	.49673
0.23	.09095	0.73	.26731	1.23	.39065	1.73	.45818	2.23	.48713	2.73	.49683
0.24	.09483	0.74	.27035	1.24	.39251	1.74	.45907	2.24	.48746	2.74	.49693
0.25	.09871	0.75	.27337	1.25	.39435	1.75	.45994	2.25	.48778	2.75	.49702
0.26	.10257	0.76	.27637	1.26	.39617	1.76	.46080	2.26	.48809	2.76	.49711
0.27	.10642	0.77	.27935	1.27	.39796	1.77	.46164	2.27	.48840	2.77	.49720
0.28	.11026	0.78	.28231	1.28	.39973	1.78	.46246	2.28	.48870	2.78	.49728
0.29	.11409	0.79	.28524	1.29	.40148	1.79	.46327	2.29	.48899	2.79	.49737
0.30	.11791	0.80	.28815	1.30	.40320	1.80	.46407	2.30	.48928	2.80	.49745
0.31	.12172	0.81	.29103	1.31	.40490	1.81	.46485	2.31	.48956	2.81	.49752
0.32	.12552	0.82	.29389	1.32	.40658	1.82	.46562	2.32	.48983	2.82	.49760
0.33	.12930	0.83	.29673	1.33	.40824	1.83	.46637	2.33	.49010	2.83	.49767
0.34	.13307	0.84	.29955	1.34	.40988	1.84	.46711	2.34	.49036	2.84	.49775
0.35	.13683	0.85	.30234	1.35	.41149	1.85	.46784	2.35	.49061	2.85	.49781
0.36	.14058	0.86	.30511	1.36	.41308	1.86	.46856	2.36	.49086	2.86	.49788
0.37	.14431	0.87	.30785	1.37	.41466	1.87	.46926	2.37	.49111	2.87	.49795
0.38	.14803	0.88	.31057	1.38	.41621	1.88	.46995	2.38	.49134	2.88	.49801
0.39	.15173	0.89	.31327	1.39	.41773	1.89	.47062	2.39	.49158	2.89	.49807
0.40	.15542	0.90	.31594	1.40	.41924	1.90	.47128	2.40	.49180	2.90	.49813
0.41	.15910	0.91	.31859	1.41	.42073	1.91	.47193	2.41	.49202	2.91	.49819
0.42	.16276	0.92	.32121	1.42	.42220	1.92	.47257	2.42	.49224	2.92	.49825
0.43	.16640	0.93	.32381	1.43	.42364	1.93	.47320	2.43	.49245	2.93	.49830
0.44	.17003	0.94	.32639	1.44	.42507	1.94	.47381	2.44	.49265	2.94	.49836
0.45	.17364	0.95	.32894	1.45	.42647	1.95	.47441	2.45	.49286	2.95	.49841
0.46	.17724	0.96	.33147	1.46	.42785	1.96	.47500	2.46	.49305	2.96	.49846
0.47	.18082	0.97	.33398	1.47	.42922	1.97	.47558	2.47	.49324	2.97	.49851
0.48	.18439	0.98	.33646	1.48	.43056	1.98	.47615	2.48	.49343	2.98	.49856
0.49	.18793	0.99	.33891	1.49	.43189	1.99	.47671	2.49	.49361	2.99	.49861

Appendix B: Normal Areas from Z to ∞

This table shows the shaded area under the curve.

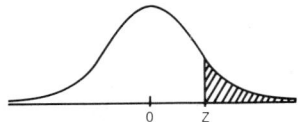

Z	Area	Z	Area	Z	Area	Z	Area	Z	Area	Z	Area
0.00	.50000	0.50	.30854	1.00	.15866	1.50	.06681	2.00	.02275	2.50	.00621
0.01	.49601	0.51	.30503	1.01	.15625	1.51	.06552	2.01	.02222	2.51	.00604
0.02	.49202	0.52	.30153	1.02	.15387	1.52	.06426	2.02	.02169	2.52	.00587
0.03	.48803	0.53	.29806	1.03	.15151	1.53	.06301	2.03	.02118	2.53	.00570
0.04	.48405	0.54	.29460	1.04	.14917	1.54	.06178	2.04	.02068	2.54	.00554
0.05	.48006	0.55	.29116	1.05	.14686	1.55	.06057	2.05	.02018	2.55	.00539
0.06	.47608	0.56	.28774	1.06	.14457	1.56	.05938	2.06	.01970	2.56	.00523
0.07	.47210	0.57	.28434	1.07	.14231	1.57	.05821	2.07	.01923	2.57	.00508
0.08	.46812	0.58	.28096	1.08	.14007	1.58	.05705	2.08	.01876	2.58	.00494
0.09	.46414	0.59	.27760	1.09	.13786	1.59	.05592	2.09	.01831	2.59	.00480
0.10	.46017	0.60	.27425	1.10	.13567	1.60	.05480	2.10	.01787	2.60	.00466
0.11	.45620	0.61	.27093	1.11	.13350	1.61	.05370	2.11	.01743	2.61	.00453
0.12	.45224	0.62	.26763	1.12	.13136	1.62	.05262	2.12	.01700	2.62	.00440
0.13	.44828	0.63	.26435	1.13	.12924	1.63	.05155	2.13	.01659	2.63	.00427
0.14	.44433	0.64	.26109	1.14	.12714	1.64	.05050	2.14	.01618	2.64	.00415
0.15	.44038	0.65	.25785	1.15	.12507	1.65	.04947	2.15	.01578	2.65	.00403
0.16	.43644	0.66	.25463	1.16	.12302	1.66	.04846	2.16	.01539	2.66	.00391
0.17	.43251	0.67	.25143	1.17	.12100	1.67	.04746	2.17	.01500	2.67	.00379
0.18	.42858	0.68	.24825	1.18	.11900	1.68	.04648	2.18	.01463	2.68	.00368
0.19	.42465	0.69	.24510	1.19	.11702	1.69	.04551	2.19	.01426	2.69	.00357
0.20	.42074	0.70	.24196	1.20	.11507	1.70	.04457	2.20	.01390	2.70	.00347
0.21	.41683	0.71	.23885	1.21	.11314	1.71	.04363	2.21	.01355	2.71	.00337
0.22	.41294	0.72	.23576	1.22	.11123	1.72	.04272	2.22	.01321	2.72	.00327
0.23	.40905	0.73	.23269	1.23	.10935	1.73	.04182	2.23	.01287	2.73	.00317
0.24	.40517	0.74	.22965	1.24	.10749	1.74	.04093	2.24	.01254	2.74	.00307
0.25	.40129	0.75	.22663	1.25	.10565	1.75	.04006	2.25	.01222	2.75	.00298
0.26	.39743	0.76	.22363	1.26	.10383	1.76	.03920	2.26	.01191	2.76	.00289
0.27	.39358	0.77	.22065	1.27	.10204	1.77	.03836	2.27	.01160	2.77	.00280
0.28	.38974	0.78	.21769	1.28	.10027	1.78	.03754	2.28	.01130	2.78	.00272
0.29	.38591	0.79	.21476	1.29	.09852	1.79	.03673	2.29	.01101	2.79	.00263
0.30	.38209	0.80	.21185	1.30	.09680	1.80	.03593	2.30	.01072	2.80	.00255
0.31	.37828	0.81	.20897	1.31	.09510	1.81	.03515	2.31	.01044	2.81	.00248
0.32	.37448	0.82	.20611	1.32	.09342	1.82	.03438	2.32	.01017	2.82	.00240
0.33	.37070	0.83	.20327	1.33	.09176	1.83	.03363	2.33	.00990	2.83	.00233
0.34	.36693	0.84	.20045	1.34	.09012	1.84	.03289	2.34	.00964	2.84	.00225
0.35	.36317	0.85	.19766	1.35	.08851	1.85	.03216	2.35	.00939	2.85	.00219
0.36	.35942	0.86	.19489	1.36	.08692	1.86	.03144	2.36	.00914	2.86	.00212
0.37	.35569	0.87	.19215	1.37	.08534	1.87	.03074	2.37	.00889	2.87	.00205
0.38	.35197	0.88	.18943	1.38	.08379	1.88	.03005	2.38	.00866	2.88	.00199
0.39	.34827	0.89	.18673	1.39	.08227	1.89	.02938	2.39	.00842	2.89	.00193
0.40	.34458	0.90	.18406	1.40	.08076	1.90	.02872	2.40	.00820	2.90	.00187
0.41	.34090	0.91	.18141	1.41	.07927	1.91	.02807	2.41	.00798	2.91	.00181
0.42	.33724	0.92	.17879	1.42	.07780	1.92	.02743	2.42	.00776	2.92	.00175
0.43	.33360	0.93	.17619	1.43	.07636	1.93	.02680	2.43	.00755	2.93	.00170
0.44	.32997	0.94	.17361	1.44	.07493	1.94	.02619	2.44	.00735	2.94	.00164
0.45	.32636	0.95	.17106	1.45	.07353	1.95	.02559	2.45	.00714	2.95	.00159
0.46	.32276	0.96	.16853	1.46	.07215	1.96	.02500	2.46	.00695	2.96	.00154
0.47	.31918	0.97	.16602	1.47	.07078	1.97	.02442	2.47	.00676	2.97	.00149
0.48	.31561	0.98	.16354	1.48	.06944	1.98	.02385	2.48	.00657	2.98	.00144
0.49	.31207	0.99	.16109	1.49	.06811	1.99	.02329	2.49	.00639	2.99	.00139

Appendix C: Critical Values for Student's t

Each entry in this table represents the critical value of t that defines the specified upper-tail area.

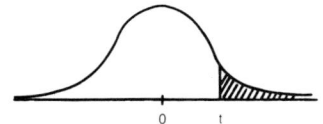

Degrees of Freedom	Upper Tail Area				
	.10	.05	.025	.01	.005
1	3.078	6.314	12.706	31.820	63.658
2	1.886	2.920	4.303	6.965	9.925
3	1.638	2.353	3.182	4.541	5.841
4	1.533	2.132	2.776	3.747	4.604
5	1.476	2.015	2.571	3.365	4.032
6	1.440	1.943	2.447	3.143	3.707
7	1.415	1.895	2.365	2.998	3.499
8	1.397	1.860	2.306	2.896	3.355
9	1.383	1.833	2.262	2.821	3.250
10	1.372	1.812	2.228	2.764	3.169
11	1.363	1.796	2.201	2.718	3.106
12	1.356	1.782	2.179	2.681	3.055
13	1.350	1.771	2.160	2.650	3.012
14	1.345	1.761	2.145	2.624	2.977
15	1.341	1.753	2.131	2.602	2.947
16	1.337	1.746	2.120	2.583	2.921
17	1.333	1.740	2.110	2.567	2.898
18	1.330	1.734	2.101	2.552	2.878
19	1.328	1.729	2.093	2.539	2.861
20	1.325	1.725	2.086	2.528	2.845
21	1.323	1.721	2.080	2.518	2.831
22	1.321	1.717	2.074	2.508	2.819
23	1.319	1.714	2.069	2.500	2.807
24	1.318	1.711	2.064	2.492	2.797
25	1.316	1.708	2.060	2.485	2.787
26	1.315	1.706	2.056	2.479	2.779
27	1.314	1.703	2.052	2.473	2.771
28	1.313	1.701	2.048	2.467	2.763
29	1.311	1.699	2.045	2.462	2.756
30	1.310	1.697	2.042	2.457	2.750
35	1.306	1.690	2.030	2.438	2.724
40	1.303	1.684	2.021	2.423	2.704
45	1.301	1.679	2.014	2.412	2.690
50	1.299	1.676	2.009	2.403	2.678
60	1.296	1.671	2.000	2.390	2.660
70	1.294	1.667	1.994	2.381	2.648
80	1.292	1.664	1.990	2.374	2.639
90	1.291	1.662	1.987	2.368	2.632
100	1.290	1.660	1.984	2.364	2.626
∞	1.282	1.645	1.960	2.326	2.576

Appendix D: Critical Values of Chi-Square

For given degrees of freedom this table shows critical values of chi-square for upper-tail shaded area.

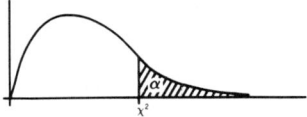

Degrees of Freedom	Upper Tail Area									
	.995	.99	.975	.95	.90	.10	.05	.025	.01	.005
1	0.000	0.000	0.001	0.004	0.016	2.706	3.841	5.024	6.635	7.880
2	0.010	0.020	0.051	0.103	0.211	4.605	5.991	7.378	9.210	10.597
3	0.072	0.115	0.216	0.352	0.584	6.251	7.815	9.349	11.345	12.838
4	0.207	0.297	0.484	0.711	1.064	7.779	9.488	11.143	13.277	14.861
5	0.412	0.554	0.831	1.145	1.610	9.236	11.071	12.833	15.086	16.750
6	0.676	0.872	1.237	1.635	2.204	10.645	12.592	14.449	16.812	18.548
7	0.989	1.239	1.690	2.167	2.833	12.017	14.067	16.013	18.476	20.278
8	1.344	1.646	2.180	2.733	3.490	13.362	15.507	17.535	20.090	21.955
9	1.735	2.088	2.700	3.325	4.168	14.684	16.919	19.023	21.666	23.590
10	2.156	2.558	3.247	3.940	4.865	15.987	18.307	20.483	23.209	25.189
11	2.603	3.053	3.816	4.575	5.578	17.275	19.675	21.920	24.725	26.757
12	3.074	3.571	4.404	5.226	6.304	18.549	21.026	23.337	26.217	28.300
13	3.565	4.107	5.009	5.892	7.042	19.812	22.362	24.736	27.688	29.820
14	4.075	4.660	5.629	6.571	7.790	21.064	23.685	26.119	29.141	31.319
15	4.601	5.229	6.262	7.261	8.547	22.307	24.996	27.488	30.578	32.801
16	5.142	5.812	6.908	7.962	9.312	23.542	26.296	28.845	32.000	34.267
17	5.697	6.408	7.564	8.672	10.085	24.769	27.587	30.191	33.409	35.718
18	6.265	7.015	8.231	9.390	10.865	25.989	28.869	31.526	34.805	37.157
19	6.844	7.633	8.907	10.117	11.651	27.204	30.144	32.852	36.191	38.582
20	7.434	8.260	9.591	10.851	12.443	28.412	31.410	34.170	37.566	39.997
21	8.034	8.897	10.283	11.591	13.240	29.615	32.671	35.479	38.932	41.401
22	8.643	9.542	10.982	12.338	14.041	30.813	33.924	36.781	40.289	42.796
23	9.260	10.196	11.689	13.091	14.848	32.007	35.172	38.076	41.638	44.181
24	9.886	10.856	12.401	13.848	15.659	33.196	36.415	39.364	42.980	45.558
25	10.520	11.524	13.120	14.611	16.473	34.382	37.652	40.646	44.314	46.928
26	11.160	12.198	13.844	15.379	17.292	35.563	38.885	41.923	45.642	48.290
27	11.808	12.878	14.573	16.151	18.114	36.741	40.113	43.195	46.963	49.645
28	12.461	13.565	15.308	16.928	18.939	37.916	41.337	44.461	48.278	50.993
29	13.121	14.256	16.047	17.708	19.768	39.087	42.557	45.722	49.588	52.336
30	13.787	14.953	16.791	18.493	20.599	40.256	43.773	46.979	50.892	53.672

Note: For chi-square above 30, compute $Z = \sqrt{2\chi^2} - \sqrt{2(\text{d.f.}) - 1}$ and use the standard normal table.

Appendix E: Critical Values of the F Distribution

For each combination of numerator and denominator degrees of freedom the table entry shows the critical value of F that corresponds to an upper-tail area of .05 (upper entry) and .01 (lower entry).

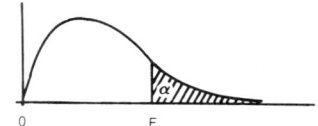

Denominator d.f.	\multicolumn{18}{c}{Numerator d.f.}

Denominator d.f.	1	2	3	4	5	6	7	8	9	10	15	20	30	40	60	120	∞
1	161	200	216	225	230	234	237	239	241	242	246	248	250	251	252	253	254
	4052	4999	5403	5625	5764	5859	5928	5981	6022	6056	6157	6209	6261	6287	6313	6339	6366
2	18.51	19.00	19.16	19.25	19.30	19.33	19.35	19.37	19.38	19.40	19.43	19.45	19.46	19.47	19.48	19.49	19.50
	98.50	99.00	99.17	99.25	99.30	99.33	99.36	99.37	99.39	99.40	99.43	99.45	99.47	99.47	99.48	99.49	99.50
3	10.13	9.55	9.28	9.12	9.01	8.94	8.89	8.85	8.81	8.79	8.70	8.66	8.62	8.59	8.57	8.55	8.53
	34.12	30.82	29.46	28.71	28.24	27.91	27.67	27.49	27.35	27.23	26.87	26.69	26.50	26.41	26.32	26.22	26.13
4	7.71	6.94	6.59	6.39	6.26	6.16	6.09	6.04	6.00	5.96	5.86	5.80	5.75	5.72	5.69	5.66	5.63
	21.20	18.00	16.69	15.98	15.52	15.21	14.98	14.80	14.66	14.55	14.20	14.02	13.84	13.75	13.65	13.56	13.46
5	6.61	5.79	5.41	5.19	5.05	4.95	4.88	4.82	4.77	4.74	4.62	4.56	4.50	4.46	4.43	4.40	4.36
	16.26	13.27	12.06	11.39	10.97	10.67	10.46	10.29	10.16	10.05	9.72	9.55	9.38	9.29	9.20	9.11	9.02
6	5.99	5.14	4.76	4.53	4.39	4.28	4.21	4.15	4.10	4.06	3.94	3.87	3.81	3.77	3.74	3.70	3.67
	13.75	10.92	9.78	9.15	8.75	8.47	8.26	8.10	7.98	7.87	7.56	7.40	7.23	7.14	7.06	6.97	6.88
7	5.59	4.74	4.35	4.12	3.97	3.87	3.79	3.73	3.68	3.64	3.51	3.44	3.38	3.34	3.30	3.27	3.23
	12.25	9.55	8.45	7.85	7.46	7.19	6.99	6.84	6.72	6.62	6.31	6.16	5.99	5.91	5.82	5.74	5.65
8	5.32	4.46	4.07	3.84	3.69	3.58	3.50	3.44	3.39	3.35	3.22	3.15	3.08	3.04	3.01	2.97	2.93
	11.26	8.65	7.59	7.01	6.63	6.37	6.18	6.03	5.91	5.82	5.52	5.36	5.20	5.12	5.03	4.95	4.86
9	5.12	4.26	3.86	3.63	3.48	3.37	3.29	3.23	3.18	3.14	3.01	2.94	2.86	2.83	2.79	2.75	2.71
	10.56	8.02	6.99	6.42	6.06	5.80	5.61	5.47	5.35	5.26	4.96	4.81	4.65	4.57	4.48	4.40	4.31
10	4.96	4.10	3.71	3.48	3.33	3.22	3.14	3.07	3.02	2.98	2.85	2.77	2.70	2.66	2.62	2.58	2.54
	10.04	7.56	6.55	5.99	5.64	5.39	5.20	5.06	4.94	4.85	4.56	4.41	4.25	4.17	4.08	4.00	3.91
15	4.54	3.68	3.29	3.06	2.90	2.79	2.71	2.64	2.59	2.54	2.40	2.33	2.25	2.20	2.16	2.11	2.07
	8.68	6.36	5.42	4.89	4.56	4.32	4.14	4.00	3.89	3.80	3.52	3.37	3.21	3.13	3.05	2.96	2.87
20	4.35	3.49	3.10	2.87	2.71	2.60	2.51	2.45	2.39	2.35	2.20	2.12	2.04	1.99	1.95	1.90	1.84
	8.10	5.85	4.94	4.43	4.10	3.87	3.70	3.56	3.46	3.37	3.09	2.94	2.78	2.69	2.61	2.52	2.42
30	4.17	3.32	2.92	2.69	2.53	2.42	2.33	2.27	2.21	2.16	2.01	1.93	1.84	1.79	1.74	1.68	1.62
	7.56	5.39	4.51	4.02	3.70	3.47	3.30	3.17	3.07	2.98	2.70	2.55	2.39	2.30	2.21	2.11	2.01
40	4.08	3.23	2.84	2.61	2.45	2.34	2.25	2.18	2.12	2.08	1.92	1.84	1.74	1.69	1.64	1.58	1.51
	7.31	5.18	4.31	3.83	3.51	3.29	3.12	2.99	2.89	2.80	2.52	2.37	2.20	2.11	2.02	1.92	1.80
60	4.00	3.15	2.76	2.53	2.37	2.25	2.17	2.10	2.04	1.99	1.84	1.75	1.65	1.59	1.53	1.47	1.39
	7.08	4.98	4.13	3.65	3.34	3.12	2.95	2.82	2.72	2.63	2.35	2.20	2.03	1.94	1.84	1.73	1.60
120	3.92	3.07	2.68	2.45	2.29	2.17	2.09	2.02	1.96	1.91	1.75	1.66	1.55	1.50	1.43	1.35	1.25
	6.85	4.79	3.95	3.48	3.17	2.96	2.79	2.66	2.56	2.47	2.19	2.03	1.86	1.76	1.66	1.53	1.38
∞	3.84	3.00	2.60	2.37	2.21	2.10	2.01	1.94	1.88	1.83	1.67	1.57	1.46	1.39	1.32	1.22	1.00
	6.63	4.61	3.78	3.32	3.02	2.80	2.64	2.51	2.41	2.32	2.04	1.88	1.70	1.59	1.47	1.32	1.00

Appendix F: Limits for Sample Skewness Coefficient

When random samples are drawn from a normal population, the sample skewness coefficient will fall within the ranges shown in the table about 90 percent of the time. If your sample skewness coefficient lies outside the specified range, it is appropriate to conclude that the population is significantly skewed, using the decision rule shown.

```
  <----                                  ---->
  Population        Population        Population
   Skewed           May Be             Skewed
    Left            Symmetric           Right

          Lower         0        Upper
          Limit                  Limit
```

Sample Size	Lower Limit	Upper Limit
25	-.711	+.711
30	-.662	+.662
35	-.621	+.621
40	-.587	+.587
50	-.534	+.534
75	-.446	+.446
100	-.389	+.389
150	-.321	+.321
200	-.280	+.280
300	-.230	+.230
400	-.200	+.200
500	-.179	+.179

Appendix G: Limits for Sample Kurtosis Coefficient

When random samples are drawn from a normal population, the sample kurtosis coefficient will fall within the ranges shown in the table about 90 percent of the time. If your sample kurtosis skewness coefficient lies outside the specified range, it is appropriate to conclude that the population is significantly different from a normal bell-shaped curve, using the decision rule shown. If the sample size is below 50, inferences about population kurtosis are risky.

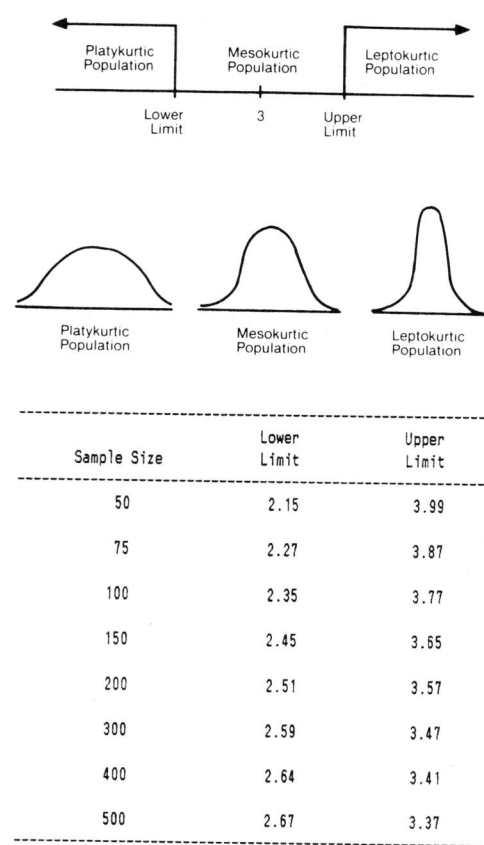

Sample Size	Lower Limit	Upper Limit
50	2.15	3.99
75	2.27	3.87
100	2.35	3.77
150	2.45	3.65
200	2.51	3.57
300	2.59	3.47
400	2.64	3.41
500	2.67	3.37

Appendix H: State Data Files

Case	STATE	ABORT	AFDC	AGE	BEDS	BIRTH	BLACK	CANCER	CARS	COLLEGE
1	ALAB	283	110	29.3	655	16.6	25.6	181	543	12.6
2	ALAS	202	359	26.1	419	23.0	3.4	70	387	22.4
3	ARIZ	185	174	29.2	444	19.7	2.8	164	485	16.8
4	ARK	161	145	30.6	562	16.7	16.3	196	442	9.7
5	CAL	494	399	29.9	492	17.4	7.7	169	541	19.8
6	COLO	354	239	28.6	527	17.8	3.5	126	608	23.0
7	CONN	435	358	32.0	595	10.9	7.0	210	662	21.2
8	DEL	422	227	29.7	705	16.5	16.1	188	547	16.3
9	FLA	466	175	34.7	597	14.3	13.8	243	651	14.7
10	GA	361	133	28.7	577	18.4	26.8	152	554	15.3
11	HAW	357	386	28.4	409	19.5	1.8	118	533	20.3
12	IDA	127	258	27.6	402	21.3	0.3	137	506	16.1
13	ILL	355	277	29.9	644	16.7	14.7	187	659	14.5
14	IND	187	203	29.2	591	16.2	7.6	182	536	12.4
15	IOWA	143	307	30.0	736	16.5	1.4	192	603	14.1
16	KANS	290	271	30.1	784	16.5	5.3	178	588	15.7
17	KENT	123	177	29.1	510	17.2	7.1	183	501	11.0
18	LA	201	148	27.4	608	19.3	29.4	165	468	13.4
19	ME	244	233	30.4	630	14.6	0.3	215	471	14.0
20	MD	464	228	30.3	596	12.6	22.7	185	562	19.8
21	MASS	623	341	31.2	774	12.7	3.9	214	579	20.0
22	MICH	313	379	28.9	543	15.5	12.9	172	566	15.2
23	MINN	279	336	29.2	759	16.6	1.3	170	602	16.7
24	MISS	96	88	27.7	657	19.5	35.2	175	462	13.0
25	MO	242	217	30.9	711	16.3	10.5	202	498	14.0
26	MONT	244	228	29.0	674	17.7	0.2	164	746	17.3
27	NEB	222	274	29.7	761	17.6	3.1	189	556	16.1
28	NEV	536	207	30.3	425	17.9	6.4	153	523	15.1
29	N H	251	271	30.1	513	15.2	0.4	191	588	18.4
30	N J	256	312	32.2	593	12.4	12.6	211	586	18.6
31	N M	213	185	27.4	477	20.0	1.8	130	521	17.3
32	N Y	629	371	31.9	745	13.2	13.7	213	408	18.7
33	N C	379	164	29.6	575	15.0	22.4	165	593	13.4
34	N D	221	277	28.3	908	19.7	0.4	161	560	15.2
35	OHIO	250	250	29.9	589	15.8	10.0	195	579	14.8
36	OKLA	216	250	30.1	592	17.2	6.8	182	607	15.7
37	ORE	365	318	30.2	451	17.1	1.4	172	574	17.2
38	PENN	415	297	32.1	727	13.7	8.8	220	489	13.8
39	R I	545	325	31.8	644	13.6	2.9	224	542	15.3
40	S C	236	107	28.2	547	16.8	30.4	157	493	14.2
41	S D	99	218	28.9	812	18.8	0.3	179	542	14.2
42	TENN	337	113	30.1	692	16.5	15.8	180	516	11.9
43	TEX	309	109	28.2	577	19.6	12.0	148	516	16.0
44	UTAH	91	314	24.2	362	31.0	0.6	94	466	20.3
45	VER	415	340	29.4	582	15.3	0.2	191	508	19.5
46	VA	408	214	29.8	598	14.3	18.9	171	571	19.2
47	WASH	462	365	29.8	398	16.9	2.6	166	557	18.8
48	W VA	102	182	30.4	727	15.9	3.3	188	497	10.5
49	WISC	291	366	29.4	615	15.6	3.9	183	535	14.9
50	WYO	76	262	27.1	560	20.4	0.7	120	609	17.2

Case	STATE	DEBT	DEFENSE	DIVORCE	DOCS	EDUC	FEMLAB	HEART	HOSP	INCOME
1	ALAB	265	380	7.1	125	4.3	47.2	305	229	7488
2	ALAS	3861	2562	8.4	129	8.3	60.6	91	496	12790
3	ARIZ	35	517	7.8	178	4.5	49.8	263	337	8791
4	ARK	159	154	9.9	121	4.4	49.0	370	204	7268
5	CAL	353	745	5.8	230	4.0	53.8	280	429	10938
6	COLO	159	446	6.4	198	4.9	59.5	222	287	10025
7	CONN	1248	1260	3.8	249	3.8	56.0	331	328	11720
8	DEL	1755	353	4.0	161	4.9	52.1	343	283	10339
9	FLA	270	380	7.8	179	3.6	45.8	399	273	8996
10	GA	257	432	6.4	143	4.1	53.1	289	249	8073
11	HAW	1932	1387	4.7	207	4.1	56.7	160	298	10101
12	IDA	347	147	7.3	107	4.5	51.8	251	246	8056
13	ILL	550	144	4.5	183	4.1	52.6	375	312	10521
14	IND	111	292	7.5	128	4.0	53.2	327	248	8936
15	IOWA	131	122	4.0	126	4.7	52.9	369	225	9358
16	KANS	185	447	5.6	157	3.8	56.1	356	234	9983
17	KENT	829	269	4.8	133	4.1	50.0	365	214	7613
18	LA	708	202	3.3	155	4.2	46.6	309	259	8458
19	ME	649	480	5.7	150	4.9	50.3	368	261	7925
20	MO	831	757	3.9	268	4.5	57.1	309	296	10460
21	MASS	1008	602	2.9	272	5.1	54.1	382	358	10125
22	MICH	315	247	4.4	158	5.4	50.4	322	313	9950
23	MINN	508	327	3.7	189	5.0	59.1	309	228	9724
24	MISS	323	481	5.5	107	5.0	48.5	327	194	6580
25	MO	207	707	5.7	163	3.7	50.6	383	257	8982
26	MONT	393	198	6.3	133	5.7	51.9	297	184	8536
27	NEB	127	242	4.1	149	4.3	54.5	350	215	9365
28	NEV	661	363	18.6	138	3.8	59.1	242	388	10727
29	N H	976	539	5.8	162	4.0	57.3	313	247	9131
30	N J	886	291	3.5	190	5.1	52.3	397	241	10924
31	N M	544	556	8.1	153	5.6	46.9	178	304	7841
32	N Y	1346	302	3.1	267	5.4	47.7	426	303	10260
33	N C	215	341	5.0	153	4.6	55.6	299	212	7819
34	N D	336	254	3.3	137	4.2	51.7	339	191	8747
35	OHIO	372	217	5.4	162	4.1	50.5	360	274	9462
36	OKLA	504	447	8.2	128	4.4	48.7	354	264	9116
37	ORE	1856	102	7.0	181	5.1	53.8	292	326	9317
38	PENN	535	313	3.0	187	4.5	46.6	428	273	9434
39	R I	1545	419	3.9	211	4.5	54.7	405	307	9444
40	S C	621	443	4.7	135	4.9	51.2	290	213	7266
41	S D	1035	209	4.1	112	4.6	55.8	374	208	7806
42	TENN	306	183	6.8	157	4.1	49.5	326	224	7720
43	TEX	174	567	7.1	152	4.2	52.3	265	252	9545
44	UTAH	367	577	5.6	162	5.2	51.6	196	314	7649
45	VER	280	248	5.0	212	5.3	54.8	343	218	7827
46	VA	360	1157	4.5	175	4.2	54.5	294	243	9392
47	WASH	388	683	7.0	180	4.8	52.3	273	307	10309
48	W VA	932	63	5.3	134	4.9	39.1	405	226	7800
49	WISC	520	122	3.7	158	4.7	56.3	353	249	9348
50	WYO	770	273	8.5	113	5.1	55.5	236	290	10898

Case	STATE	INFMOR	LIFER	NEAST	OVR65	PCRIME	POP	POPCH	POPDEN	PUBAID
1	ALAB	16.1	69.8	0	11.3	1527	3.9	12.9	77	7.6
2	ALAS	14.4	20.9	0	2.9	1386	0.4	32.4	1	4.8
3	ARIZ	13.1	20.9	0	11.3	2155	2.7	53.1	24	3.3
4	ARK	16.4	53.5	0	13.7	1119	2.3	18.8	44	7.0
5	CAL	11.8	33.3	0	10.2	2317	23.7	18.5	151	9.3
6	COLO	11.2	31.3	0	8.6	2031	2.9	30.7	28	3.7
7	CONN	11.6	51.2	1	11.7	1701	3.1	2.5	638	5.3
8	DEL	13.2	43.0	1	10.0	1631	0.6	8.6	308	6.6
9	FLA	14.1	22.3	0	17.3	2507	9.7	43.4	180	4.4
10	GA	15.4	62.9	0	9.5	1699	5.5	19.1	94	7.0
11	HAW	11.1	55.0	0	7.9	1847	1.0	25.3	150	7.3
12	IDA	11.7	36.9	0	9.9	1239	0.9	32.4	12	2.8
13	ILL	15.7	61.1	0	11.0	1243	11.4	2.8	205	7.4
14	IND	13.1	61.3	0	10.7	1313	5.5	5.7	153	3.9
15	IOWA	12.6	67.4	0	13.3	1669	2.9	3.1	52	4.5
16	KANS	12.5	55.6	0	13.0	1521	2.4	5.1	29	4.0
17	KENT	12.7	71.8	0	11.2	1041	3.7	13.7	92	7.4
18	LA	17.3	70.7	0	9.6	1524	4.2	15.3	95	8.0
19	ME	10.4	62.3	1	12.5	1183	1.1	13.2	36	6.8
20	MD	14.7	45.7	1	9.4	1698	4.2	7.5	429	6.2
21	MASS	11.1	65.7	1	12.7	1740	5.7	0.8	733	8.0
22	MICH	13.8	64.9	0	9.8	1741	9.3	4.2	163	9.5
23	MINN	12.0	64.2	0	11.8	1246	4.1	7.1	51	4.4
24	MISS	18.7	69.1	0	11.5	1179	2.5	13.7	53	11.3
25	MO	14.8	62.2	0	13.2	1669	4.9	5.1	71	6.1
26	MONT	11.6	47.3	0	10.7	951	0.8	13.3	5	3.5
27	NEB	13.0	60.6	0	13.1	915	1.6	5.7	21	3.3
28	NEV	12.5	13.1	0	8.2	2907	0.8	63.5	7	2.6
29	N H	10.4	43.0	1	11.2	1313	0.9	24.8	102	3.1
30	N J	13.0	51.8	1	11.7	1878	7.4	2.7	986	7.5
31	N M	14.1	43.8	0	8.9	1493	1.3	27.8	11	6.1
32	N Y	14.0	62.4	1	12.3	2062	17.6	-3.8	371	8.3
33	N C	16.6	70.3	0	10.3	1423	5.9	15.5	120	5.7
34	N D	13.5	64.8	0	12.3	488	0.7	5.6	9	3.0
35	OHIO	13.3	63.8	0	10.8	1466	10.8	1.3	263	6.7
36	OKLA	14.3	52.2	0	12.4	1693	3.0	18.2	44	5.0
37	ORE	12.9	32.0	0	11.5	1748	2.6	25.9	27	4.3
38	PENN	13.7	74.9	1	12.9	1039	11.9	0.6	264	6.8
39	R I	13.6	62.2	1	13.4	1716	0.9	-0.3	898	7.3
40	S C	18.6	69.0	0	9.2	1670	3.1	20.4	103	7.6
41	S D	13.5	63.7	0	13.2	693	0.7	3.6	9	3.8
42	TENN	14.8	64.4	0	11.3	1501	4.6	16.9	112	6.6
43	TEX	14.3	59.1	0	9.6	1853	14.2	27.1	54	3.9
44	UTAH	11.4	52.5	0	7.5	1322	1.5	37.9	18	3.4
45	VER	10.4	54.8	1	11.4	1527	0.5	15.0	55	6.6
46	VA	13.8	52.0	0	9.5	1203	5.3	14.9	135	4.7
47	WASH	12.5	36.8	0	10.4	1862	4.1	21.0	62	4.3
48	W VA	15.1	70.6	0	12.2	738	2.0	11.8	81	6.3
49	WISC	11.2	68.7	0	12.0	1079	4.7	6.5	87	6.6
50	WYO	13.0	33.9	0	7.9	904	0.5	41.6	5	1.8

Case	STATE	PUPIL	REPUB	SALETX	SPANISH	SPEND	SEAST	TDEATH	UNBEN	UND25
1	ALAB	23.6	48.7	4.000	0.9	1326	1	30.2	12.0	21.6
2	ALAS	17.3	54.3	0.000	2.4	6257	0	20.4	17.8	23.5
3	ARIZ	19.4	60.6	4.000	16.2	1547	0	41.0	12.5	21.7
4	ARK	19.0	48.1	3.000	0.8	1200	1	32.3	12.1	21.0
5	CAL	20.8	52.7	4.750	19.2	1834	0	26.5	15.7	20.4
6	COLO	18.6	55.1	3.000	11.8	1578	0	25.3	10.8	22.5
7	CONN	16.1	48.2	7.500	4.0	1583	0	18.8	12.5	20.1
8	DEL	17.3	47.2	0.000	1.6	1809	0	19.9	13.7	21.8
9	FLA	21.0	55.5	4.000	8.8	1309	1	28.4	12.0	19.0
10	GA	20.6	41.0	3.000	1.1	1366	1	31.5	9.9	21.8
11	HAW	23.4	42.9	4.000	7.4	1945	0	23.3	13.5	19.7
12	IDA	20.9	66.5	3.000	3.9	1366	0	39.3	12.4	25.5
13	ILL	19.1	49.6	4.000	5.6	1587	0	19.9	18.6	22.2
14	IND	20.3	56.0	4.000	1.6	1243	0	25.0	13.2	22.1
15	IOWA	16.7	51.3	3.000	0.9	1686	0	23.6	11.9	23.1
16	KANS	16.2	57.8	3.000	2.7	1586	0	24.4	12.8	22.7
17	KENT	20.5	49.0	5.000	0.7	1477	1	26.6	15.0	23.0
18	LA	17.9	51.2	3.000	2.4	1560	1	30.5	15.1	23.0
19	ME	21.2	45.6	5.000	0.4	1405	0	24.3	10.8	20.0
20	MD	18.6	44.2	5.000	1.5	1809	0	17.5	14.1	20.8
21	MASS	14.9	41.9	5.000	2.5	1796	0	17.4	15.0	20.1
22	MICH	21.5	49.0	4.000	1.8	1880	0	21.6	18.1	24.4
23	MINN	17.6	42.5	4.000	0.8	1894	0	23.0	15.3	21.7
24	MISS	18.7	49.5	5.000	1.0	1354	1	32.3	12.8	23.1
25	MO	17.9	51.2	3.125	1.1	1280	0	24.5	13.4	22.4
26	MONT	16.6	56.8	0.000	1.3	1769	0	39.6	12.7	22.4
27	NEB	15.7	65.6	3.000	1.8	1544	0	22.4	11.3	23.7
28	NEV	21.1	63.6	5.750	6.7	1867	0	45.2	12.0	21.2
29	N H	18.1	57.7	0.000	0.6	1340	0	21.1	8.1	20.7
30	N J	16.6	52.0	5.000	6.7	1688	0	16.8	16.1	18.5
31	N M	19.5	55.0	3.750	36.6	1658	0	48.3	15.0	25.0
32	N Y	18.7	46.7	4.000	9.5	2204	0	14.9	20.0	16.9
33	N C	20.4	49.3	3.000	1.0	1301	1	27.6	9.4	22.8
34	N D	15.6	64.3	3.000	0.6	1840	0	21.3	13.4	24.8
35	OHIO	20.0	51.5	4.000	1.1	1431	0	21.4	16.2	19.7
36	OKLA	17.6	60.5	2.000	1.9	1406	0	32.2	11.7	21.6
37	ORE	18.9	48.3	0.000	2.5	1899	0	28.7	13.7	19.6
38	PENN	17.6	49.6	6.000	1.3	1469	0	19.7	16.7	19.4
39	R I	16.7	37.3	6.000	2.1	1754	0	15.8	14.0	20.8
40	S C	20.9	49.5	4.000	1.1	1269	1	31.1	9.4	22.6
41	S D	16.5	60.5	4.000	0.6	1586	0	32.9	11.2	24.9
42	TENN	21.1	48.7	4.500	0.7	1291	1	29.1	13.8	20.7
43	TEX	18.7	55.3	4.000	21.0	1362	0	31.0	12.3	21.7
44	UTAH	26.6	72.8	4.000	4.1	1615	0	24.7	13.7	26.2
45	VER	14.8	44.4	3.000	0.6	1600	0	30.4	13.7	20.9
46	VA	18.1	53.0	3.000	1.5	1440	1	19.3	12.1	22.0
47	WASH	21.5	49.6	4.500	2.9	1782	0	28.2	14.6	20.1
48	W VA	19.2	45.3	3.000	0.7	1521	1	29.2	13.2	19.4
49	WISC	17.8	47.9	4.000	1.3	1799	0	21.9	14.2	22.6
50	WYO	18.3	62.6	3.000	5.2	2335	0	46.2	10.8	25.1

Case	STATE	UNEM	UNION	URBAN	VCRIME	VOTE	WAIT	WEST
1	ALAB	8.8	21.8	62.0	273	49.0	0	0
2	ALAS	9.5	33.7	43.2	317	58.3	1	1
3	ARIZ	6.6	15.8	75.0	402	45.2	0	1
4	ARK	7.6	16.0	39.1	218	51.6	1	0
5	CAL	6.8	27.0	94.9	437	49.5	0	1
6	COLO	5.6	18.1	80.9	309	56.8	0	1
7	CONN	5.9	22.9	88.3	168	61.2	1	0
8	DEL	7.7	25.1	67.0	307	54.9	0	0
9	FLA	6.0	11.7	87.9	557	49.6	1	0
10	GA	6.4	15.0	60.0	300	41.7	1	0
11	HAW	5.0	27.9	79.1	66	43.6	0	1
12	IDA	7.9	18.5	18.3	241	68.5	0	1
13	ILL	8.3	30.6	81.0	240	57.8	0	0
14	IND	9.6	30.4	69.8	194	57.7	1	0
15	IOWA	5.7	22.0	40.1	287	62.9	1	0
16	KANS	4.4	15.4	46.8	238	57.0	1	0
17	KENT	8.1	24.0	44.5	143	50.0	1	0
18	LA	6.7	16.3	63.4	408	53.7	0	0
19	ME	7.7	24.2	33.0	147	64.8	1	0
20	MD	6.4	22.6	88.8	410	50.2	1	0
21	MASS	5.6	24.9	85.3	334	59.3	1	0
22	MICH	12.6	37.4	82.7	339	59.8	1	0
23	MINN	5.7	26.2	64.6	103	70.4	1	0
24	MISS	7.5	16.3	27.1	222	52.1	0	0
25	MO	7.0	27.6	65.3	287	58.9	1	0
26	MONT	6.0	29.3	24.0	164	65.2	0	1
27	NEB	4.0	18.2	44.2	115	56.8	1	0
28	NEV	6.2	23.8	82.0	366	41.3	0	1
29	N H	4.7	15.8	50.7	118	57.8	1	0
30	N J	7.2	25.6	91.4	263	55.1	1	0
31	N M	7.4	18.9	42.4	431	51.4	0	1
32	N Y	7.5	38.7	90.1	345	48.0	0	0
33	N C	6.5	9.6	52.7	339	43.9	0	0
34	N D	4.9	17.1	35.9	36	65.1	0	0
35	OHIO	8.4	31.5	80.3	232	55.4	1	0
36	OKLA	4.8	15.3	58.5	268	52.8	0	0
37	ORE	8.2	26.0	64.9	291	61.6	1	1
38	PENN	7.8	34.6	81.9	156	52.0	1	0
39	R I	7.2	28.4	92.2	268	59.0	0	0
40	S C	6.9	7.8	59.7	493	40.7	0	0
41	S D	4.7	14.7	15.9	94	67.4	0	0
42	TENN	7.2	19.1	62.8	229	48.9	0	0
43	TEX	5.2	11.4	80.0	278	45.6	0	1
44	UTAH	6.2	17.8	79.0	192	65.5	0	1
45	VER	6.4	18.0	22.3	109	58.3	0	0
46	VA	5.1	14.7	69.6	151	48.0	0	0
47	WASH	7.5	34.4	80.4	271	58.0	1	1
48	W VA	9.4	34.4	37.1	112	52.8	1	0
49	WISC	7.0	28.6	66.8	94	67.7	1	0
50	WYO	3.9	18.6	15.3	313	54.2	0	1

Appendix I: Time-Series Data Files

YEAR	GNP	C	I	G	X	M	X-M
1960	515.3	330.7	78.2	100.6	29.9	24.0	5.9
1961	533.8	341.1	77.1	108.4	31.1	23.9	7.2
1962	574.6	361.9	87.6	118.2	33.1	26.2	6.9
1963	606.9	381.7	93.1	123.8	35.7	27.5	8.2
1964	649.8	409.3	99.6	130.0	40.5	29.6	10.9
1965	705.1	440.7	116.2	138.6	42.9	33.2	9.7
1966	772.0	477.3	128.6	158.6	46.6	39.1	7.5
1967	816.4	503.6	125.7	179.7	49.5	42.1	7.4
1968	892.7	552.5	137.0	197.7	54.8	49.3	5.5
1969	963.9	597.9	153.2	207.3	60.4	54.7	5.6
1970	1015.5	640.0	148.8	218.2	68.9	60.5	8.5
1971	1102.7	691.6	172.5	232.4	72.4	66.1	6.3
1972	1212.8	757.6	202.0	250.0	81.4	78.2	3.2
1973	1359.3	837.2	238.8	266.5	114.1	97.3	16.8
1974	1472.8	916.5	240.8	299.1	151.5	135.2	16.3
1975	1598.4	1012.8	219.6	335.0	161.3	130.3	31.1
1976	1782.8	1129.3	277.7	356.9	177.7	158.9	18.8
1977	1990.5	1257.2	344.1	387.3	191.6	189.7	1.9
1978	2249.7	1403.5	416.8	425.2	227.5	223.4	4.1
1979	2508.2	1566.8	454.8	467.8	291.2	272.5	18.8
1980	2732.0	1732.6	437.0	530.3	351.0	318.9	32.1
1981	3052.6	1915.1	515.5	588.1	382.8	348.9	33.9
1982	3166.0	2050.7	447.3	641.7	361.9	335.6	26.3
1983	3401.6	2229.3	501.9	675.7	354.1	359.4	-5.3
1984	3774.7	2423.0	674.0	736.8	384.6	443.8	-59.2
1985	3992.5	2581.9	670.4	814.6	370.4	444.8	-74.4

YEAR	PI	PT	PS	DPI	APS	DPI-CAP	DEFICIT
1960	409.4	50.5	20.8	358.9	5.8	1986	-3.0
1961	426.0	52.2	24.9	373.8	6.6	2034	3.9
1962	453.2	57.0	25.9	396.2	6.5	2123	4.2
1963	476.3	60.5	24.6	415.8	5.9	2197	-0.3
1964	510.2	58.8	31.5	451.4	7.0	2352	3.3
1965	552.0	65.2	34.3	486.8	7.0	2505	-0.5
1966	600.8	74.9	36.0	525.9	6.8	2675	1.8
1967	644.5	82.4	45.1	562.1	8.0	2828	13.2
1968	707.2	97.7	42.5	609.6	7.0	3037	6.0
1969	772.9	116.3	42.2	656.7	6.4	3239	-8.4
1970	831.8	116.2	57.7	715.6	8.1	3489	12.4
1971	894.0	117.3	66.3	776.8	8.5	3740	22.0
1972	981.6	142.0	61.4	839.6	7.3	4000	16.8
1973	1101.7	152.0	89.0	949.8	9.4	4481	5.6
1974	1210.1	171.8	96.7	1038.4	9.3	4855	11.6
1975	1313.4	170.6	104.6	1142.8	9.2	5291	69.4
1976	1451.4	198.7	95.8	1252.6	7.6	5744	53.5
1977	1607.5	228.1	90.7	1379.3	6.6	6262	46.0
1978	1812.4	261.1	110.2	1551.2	7.1	6968	29.3
1979	2034.0	304.7	118.1	1729.3	6.8	7682	16.1
1980	2258.5	340.5	136.9	1918.0	7.1	8422	61.3
1981	2520.9	393.3	159.4	2127.6	7.5	9247	63.8
1982	2670.8	409.3	153.9	2261.4	6.8	9732	145.9
1983	2836.4	411.1	133.2	2425.4	5.5	10339	179.4
1984	3111.9	441.8	172.5	2670.2	6.5	11279	172.9
1985	3294.2	493.1	129.7	2801.1	4.6	11727	197.3

YEAR	LABOR	UNEMPT	PR-MEN	PR-FEM	PROD	FARM	COMP
1960	69.6	5.5	86.0	37.6	67.5	37	33.6
1961	70.5	6.7	85.7	38.0	69.9	39	34.9
1962	70.6	5.5	84.8	37.8	72.4	41	36.5
1963	71.8	5.7	84.4	38.3	75.3	45	37.9
1964	73.1	5.2	84.2	38.9	78.5	47	39.9
1965	74.5	4.5	83.9	39.4	80.9	52	41.4
1966	75.8	3.8	83.6	40.1	83.1	53	44.3
1967	77.3	3.8	83.4	41.1	85.3	58	46.7
1968	78.7	3.6	83.1	41.6	87.6	62	50.3
1969	80.7	3.5	82.8	42.7	87.7	63	53.8
1970	82.8	4.9	82.6	43.3	88.3	66	57.7
1971	84.4	5.9	82.1	43.3	91.2	74	61.5
1972	87.0	5.6	81.6	43.7	94.1	78	65.5
1973	89.4	4.9	81.3	44.4	95.9	81	70.9
1974	91.9	5.6	81.0	45.3	93.9	79	77.6
1975	93.8	8.5	80.3	46.0	95.7	89	85.2
1976	96.2	7.7	79.8	47.0	98.3	94	92.8
1977	99.0	7.1	79.7	48.1	100.0	100	100.0
1978	102.3	6.1	79.8	49.6	100.8	108	108.5
1979	105.0	5.8	79.8	50.6	99.6	119	119.1
1980	106.9	7.1	79.4	51.3	99.2	112	131.5
1981	108.7	7.6	79.0	52.1	100.7	131	143.7
1982	110.2	9.7	78.7	52.7	100.3	133	154.9
1983	111.6	9.6	78.5	53.1	102.9	120	161.5
1984	113.5	7.5	78.3	53.7	105.0	139	167.8
1985	115.5	7.2	78.1	54.7	105.3	139	174.5

YEAR	FRB	CPI	PPI	FPI	CH-CPI	R-3MO	R-BOND	PRIME
1960	80.1	88.7	93.7	52	1.5	2.93	4.41	4.82
1961	77.3	89.6	93.7	53	0.7	2.38	4.35	4.50
1962	81.4	90.6	94.0	53	1.2	2.78	4.33	4.50
1963	83.5	91.7	93.7	53	1.6	3.16	4.26	4.50
1964	85.6	92.9	94.1	52	1.2	3.55	4.40	4.50
1965	89.5	94.5	95.7	54	1.9	3.95	4.49	4.54
1966	91.1	97.2	98.8	58	3.4	4.88	5.13	5.63
1967	86.7	100.0	100.0	55	3.0	4.32	5.51	5.61
1968	87.0	104.2	102.8	56	4.7	5.34	6.18	6.30
1969	86.7	109.8	106.6	59	6.1	6.68	7.03	7.96
1970	79.2	116.3	110.3	60	5.5	6.46	8.04	7.91
1971	77.4	121.3	113.7	62	3.4	4.35	7.39	5.72
1972	82.8	125.3	117.2	69	3.4	4.07	7.21	5.25
1973	87.0	133.1	127.9	98	8.8	7.04	7.44	8.03
1974	82.6	147.7	147.5	105	12.2	7.89	8.57	10.81
1975	72.3	161.2	163.4	101	7.0	5.84	8.83	7.86
1976	77.4	170.5	170.6	102	4.8	4.99	8.43	6.84
1977	81.4	181.5	181.7	100	6.8	5.27	8.02	6.83
1978	84.2	195.4	195.9	115	9.0	7.22	8.73	9.06
1979	84.6	217.4	217.7	132	13.3	10.04	9.63	12.67
1980	79.3	246.8	247.0	134	12.4	11.51	11.94	15.27
1981	78.3	272.4	269.8	139	8.9	14.03	14.17	18.87
1982	70.3	289.1	280.7	133	3.9	10.69	13.79	14.86
1983	74.0	298.4	285.2	134	3.8	8.63	12.04	10.79
1984	80.8	311.1	291.1	142	4.0	9.58	12.71	12.04
1985	80.3	322.2	293.8	128	3.8	7.48	11.37	9.93

YEAR	DJIA	M1	M2	M3	CH-M1	CDEBT	FDEBT	MDEBT
1960	618.04	141.8	312.3	315.3	0.6	44.3	290.9	207.5
1961	691.55	146.5	335.5	341.0	3.3	45.4	292.9	228.0
1962	639.76	149.2	362.7	371.4	1.8	50.4	303.3	251.4
1963	714.81	154.7	393.2	406.0	3.7	57.1	310.8	278.5
1964	834.05	161.9	424.8	442.5	4.7	64.7	316.8	305.9
1965	910.88	169.5	459.4	482.2	4.7	72.8	323.2	333.3
1966	873.60	173.7	480.0	505.1	2.5	78.2	329.5	356.5
1967	879.12	185.1	524.3	557.1	6.6	81.8	341.3	381.2
1968	906.00	199.4	566.3	606.2	7.7	90.1	369.8	410.9
1969	876.72	205.8	589.5	615.0	3.2	99.4	367.1	441.4
1970	753.19	216.6	628.2	677.5	5.2	103.9	382.6	473.5
1971	884.76	230.8	712.8	776.2	6.6	116.4	409.5	524.0
1972	950.71	252.0	805.2	886.0	9.2	131.3	437.3	597.1
1973	923.88	265.9	861.0	985.0	5.5	152.9	468.4	672.3
1974	759.37	277.5	908.4	1070.4	4.4	162.2	486.2	732.3
1975	802.49	291.1	1023.1	1172.2	4.9	169.4	544.1	791.7
1976	974.92	310.3	1163.6	1311.8	6.6	190.7	631.9	878.5
1977	894.63	335.3	1286.6	1472.5	8.1	226.6	709.1	1009.8
1978	820.23	363.0	1388.9	1646.4	8.3	269.4	780.4	1161.9
1979	844.40	389.0	1497.9	1803.6	7.2	307.1	833.8	1327.3
1980	891.41	414.8	1631.4	1988.5	6.6	296.3	914.3	1457.5
1981	932.92	441.8	1794.4	2235.8	6.5	312.9	1003.9	1564.0
1982	884.36	480.8	1954.9	2446.8	8.8	328.3	1147.0	1631.3
1983	1190.34	528.0	2188.8	2701.7	9.8	376.0	1381.9	1811.4
1984	1178.48	558.5	2371.7	2995.0	5.8	452.4	1576.7	2022.5
1985	1328.23	624.7	2563.6	3213.5	11.9	542.8	1827.5	2233.6

Index

Ordering Diskettes

You will need the *Exploring Statistics with the IBM PC, Second Edition,* software to use this book. If your copy does not have diskettes inside the back cover, you may order a program diskette and database diskette set for $31.95 plus postage and handling (code number 11819) from Addison-Wesley Publishing Company. This price is subject to change without notice and is the suggested list price. All diskettes are double-sided and designed for the IBM PC equipment specified in the preface. To order, call or write:

Addison-Wesley Publishing Company, Inc.
Attn.: Order Department
Reading, MA 01867

Telephone: (617) 944-3700

Site License

A site license for unrestricted use of this software, at a single installation, is available from the publisher for a moderate annual fee. The fee depends on the number of users. The site license is appropriate for a network or PC lab. Individuals would still have the option of ordering personal copies, either through bookstores or directly from the publisher. For license information, write Addison-Wesley's Marketing Department at the address indicated above.